SpringerBriefs in Applied Sciences and Technology

SpringerBriefs present concise summaries of cutting-edge research and practical applications across a wide spectrum of fields. Featuring compact volumes of 50–125 pages, the series covers a range of content from professional to academic.

Typical publications can be:

- A timely report of state-of-the art methods
- An introduction to or a manual for the application of mathematical or computer techniques
- A bridge between new research results, as published in journal articles
- A snapshot of a hot or emerging topic
- An in-depth case study
- A presentation of core concepts that students must understand in order to make independent contributions

SpringerBriefs are characterized by fast, global electronic dissemination, standard publishing contracts, standardized manuscript preparation and formatting guidelines, and expedited production schedules.

On the one hand, **SpringerBriefs in Applied Sciences and Technology** are devoted to the publication of fundamentals and applications within the different classical engineering disciplines as well as in interdisciplinary fields that recently emerged between these areas. On the other hand, as the boundary separating fundamental research and applied technology is more and more dissolving, this series is particularly open to trans-disciplinary topics between fundamental science and engineering.

Indexed by EI-Compendex, SCOPUS and Springerlink.

More information about this series at http://www.springer.com/series/8884

María José Piernas Muñoz
Elizabeth Castillo Martínez

Prussian Blue Based Batteries

 Springer

María José Piernas Muñoz
Electrochemical Energy Storage–Chemical
    Science and Engineering Division
Argonne National Laboratory
Lemont, IL
USA

Elizabeth Castillo Martínez
Department of Chemistry
University of Cambridge
Cambridge
UK

ISSN 2191-530X   ISSN 2191-5318   (electronic)
SpringerBriefs in Applied Sciences and Technology
ISBN 978-3-319-91487-9   ISBN 978-3-319-91488-6   (eBook)
https://doi.org/10.1007/978-3-319-91488-6

Library of Congress Control Number: 2018941531

Printed on acid-free paper

This Springer imprint is published by the registered company Springer International Publishing AG part of Springer Nature
The registered company address is: Gewerbestrasse 11, 6330 Cham, Switzerland

Preface

Batteries have become an essential tool for our daily life all over the world and keep gaining importance. Their ability to store energy along with their simplicity of construction and relatively low cost have fostered this process.

Lead-acid batteries, for instance, have started our cars for more than a century and still continue doing it. Recently, with the aim of decreasing the amount of CO_2 released to the atmosphere, Li-ion batteries are being implemented to replace gasoline, leading to the resurgence of electric vehicles. However, the applications of this technology do not end here. Li-ion technology is also integrated in most of the portable electronic devices we handle, such as mobile phones, laptops and tablets, making our daily life easier. Besides, it is currently being evaluated and tested for grid energy storage by coupling it with renewable sources (wind, solar), opening a window towards a cleaner and more sustainable energy landscape.

Despite the success of Li-ion batteries associated with the high energy and power density they are capable of delivering, the desire to improve these parameters and to use more abundant elements have pushed scientists to explore other options beyond Li-ion batteries. These include Na-ion, K-ion and multivalent ions (Mg^{2+}, Ca^{2+}, Al^{3+}, etc.), which are nowadays under exhaustive study. For all these batteries, that we could coin as A-ion batteries (A = Li, Na, K, Mg, Ca, Al, ...), several materials are being investigated with the objective of finding those systems that could fulfill the requirements necessary for real battery applications. Notwithstanding, this book is focused on one in particular, the so-called Prussian Blue and its derivatives, which meet many specifications to be part of the next generation grid-scale batteries.

The main motivation to write about this particular topic originates in the fascination these systems aroused in me, and in one of my then supervisors and co-author of this book, while we study them during my years of Ph.D. research.

The book is structured in five chapters. Chapter 1 is intended to provide an overview of batteries: starting with some history, continuing with the basic fundaments and concepts, and finishing with those battery technologies of greater relevance for the purpose of this book, i.e. Li-ion batteries and beyond. Chapter 2 is dedicated to the material subject of study, known as Prussian Blue, addressing a detailed description of its structure and that of its derivatives and analogues, as well as the features that characterize them and a general synopsis of the wide number of applications this material is considered for. Chapters 3 and 4 are conceived for acquiring a deeper understanding regarding the application of these materials to A-ion batteries, either in aqueous (Chap. 3) or in non-aqueous media (Chap. 4). In these chapters, several strategies to improve the properties of PB and achieve their maximum/optimal performance are described. To conclude, the main ideas extracted from previous chapters and future perspectives are summarized in Chap. 5.

We hope the reader both enjoys and learns from this book as much as we did during the writing process.

Lemont, USA   María José Piernas Muñoz
Cambridge, UK   Elizabeth Castillo Martínez

Acknowledgements

First of all, we would like to acknowledge all the scientists who have performed pioneering research in the field of Prussian Blue based batteries as well as to those who have contributed to enrich it, making this book possible.

We would also like to deeply acknowledge our editor Dr. Mayra Castro, for offering us the great challenge and opportunity of writing this book, for the infinite patience she has demonstrated with us and for her priceless help and guidance throughout the writing and reviewing process.

María José Piernas Muñoz would like to thank all the people who have contributed to the beginning and development of her scientific career, and for helping her to grow both professionally and personally. Especially to Prof. Mª Dolores Santana Lario, Prof. Gabriel García Sánchez, Prof. Teófilo Rojo Aparicio, Dr. Elizabeth Castillo Martínez, Prof. Michel Armand, Dr. Cristopher Johnson and Dr. Ira Bloom and to all the colleagues and friends with whom she has had the pleasure of working. She would also like to acknowledge the research centre CIC EnergiGUNE, for the opportunity they offered her to develop her Ph.D. in Prussian Blue type materials for battery applications, without which this book would not have been possible either. And to her family and friends for their unconditional support and encouragement.

Elizabeth Castillo Martínez would like to thank all the mentors and so many colleagues all around the world who have shared stimulating discussions. Especially to those who has been so important for her battery research and who has served her as an inspiration, to Prof. Clare P. Grey FRS '*who has opened me the door to all the exciting ongoing research in her lab where I could freely participate and learn so much about batteries*' and to Prof. Michel Armand, '*who was amazing to work with in the lab, constantly proposing new hypotheses and experiments to confirm those, and whose enthusiasm about every new observation was like that of a fresh student*'. She would also like to thank her growing family for their help and support during the writing of this book.

Contents

Chapter 1
Introduction to Batteries

1.1 The Origin of Batteries

Electrochemistry is the science dedicated to studying the relationship existent between the occurrence of certain chemical processes and the electrical current needed or generated during such chemical processes. It investigates the chemical changes, specifically the oxidation and reduction reactions (designated as redox), caused by passing electrical current, as well as the reverse phenomenon, i.e. the generation of electron movement as a result of chemical reactions.

The origin of the electrochemistry dates back to the end of the eighteenth century and is attributed to the scientific findings achieved by two Italians. Luigi Galvani, who discovered electricity indirectly, by contacting two different metals that stimulated the contraction of a leg of a dead frog [1]. And Alessandro Volta, who proposed an explanation to the previous fact developing what today is known as 'theory of contact', which states that durable flow of electric current could be generated by the mere contact of conducting substances of different kinds, and invented the first d.c. power source to demonstrate his theory, the 'Voltaic Pile', consisting of a column of alternatively packed silver and zinc plates capable to decompose water into hydrogen and oxygen (electrolysis of water) [2].

Although there are some speculations that suggest that the art of generating direct current (d.c.) was known almost 2 millenniums prior to Volta's Pile, as a result of the discovery of the so-called Baghdad candle, this has not yet been verified [3]. Therefore, Volta's Pile is the milestone considered as the starting point to lay the foundations of electrochemistry.

By improving Volta's Pile, a wide variety of electrochemical batteries were developed during the nineteenth and twentieth centuries, such as the electrical generators devised by Daniel, Planté or Leclanché, among others [4]. The principal breakthroughs achieved in the development of batteries are listed in the timeline presented in Fig. 1.1, however we will not go into more detail here.

© The Author(s) 2018
M. J. Piernas Muñoz and E. Castillo Martínez, *Prussian Blue Based Batteries*,
SpringerBriefs in Applied Sciences and Technology,
https://doi.org/10.1007/978-3-319-91488-6_1

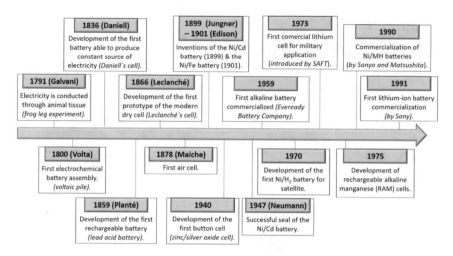

Fig. 1.1 Chronology of the major events in the history of batteries. Information extracted from [5]

1.2 Battery Definition and Working Principle

A battery is a device capable of converting the chemical energy, contained in the active materials that compose it, into electric energy by electrochemical redox reactions. Although 'battery' is the term generally adopted to refer to them, the basic electrochemical unit is denominated 'cell'. Indeed, a battery is built by connecting one or more cells in series or in parallel, what allows to obtain a higher output voltage or capacity [6].

The basic components of an electrochemical cell are two electrodes (an anode and a cathode), the electrolyte and a cell container:

- The **anode** (or negative electrode) is usually made of materials with very few electrons in their valence shell, as metals or compounds that include them.
- The **cathode** (or positive electrode) is commonly made of materials that have nearly full valence shells, such as compounds that include oxygen, chlorine or both.
- The **electrolyte** is the ionic conductor that facilitates the movement of ions between the electrodes. Electrolytes can be liquid (aqueous (either acids or alkaline) or an organic solvent, containing a salt in solution), ceramic or polymeric. In most of the cases, when a liquid electrolyte is used, a separator which provides electronic insulation between the anode and the cathode while allowing ionic transport between them is also required.
- The **cell container** is the casing where the cell is assembled. It must be inert to avoid undesired side reactions with the electrodes and electrolyte and ensure a good sealing.

The operating principle of a battery can be described as detailed below. When the anode is connected to the cathode through an external circuit, the cell undergoes discharge spontaneously. During discharge, the anode material releases electrons (is oxidized) and the cathode accepts them (is reduced). That is, redox reactions occur and electrons flow from the anode to the cathode through an external circuit. Simultaneously, the ions resulting from the dissociation of the electrolyte travel to the proximities of the electrodes to compensate the charges formed there: the cations going to the negatively polarized electrode (cathode) and the anions to the positively polarized electrode (anode). For rechargeable batteries, by applying a voltage or a current from an external source, the redox processes are reversed and the cell gets charged [7, 8]. In the particular case of A-ion batteries (A = alkali, alkali-earth or Al) and as a result of the redox reactions, 'A' ions are extracted from one of the electrodes (from the anode if the cell is discharged) and travel through the electrolyte to be inserted in the other electrode to also maintain electrical neutrality (see Sect. 1.3).

The performance of an electrochemical cell mainly judged by two parameters: the **voltage** and the **capacity** it is capable of delivering. These two factors are mostly determined by the chemistry of the active materials that form the electrodes. Therefore, the selection of the anode and cathode materials is crucial. Among the possible anode-cathode combinations employed to construct a cell, those that provide a high voltage and capacity are generally preferred in A-ion batteries [6], although certainly these parameters (voltage and capacity) will define the applications in which the cells will be used [6]. **Cell voltage** (expressed in volts, V) depends on the potential of the electrodes and its theoretical value is given by the difference in the 'standard reduction potential' between the two materials/electrodes [6]. Although experimentally there might be differences depending on several parameters such as the electrolyte. **Capacity** is the electric charge that a battery can deliver over time at a certain discharge current (C-rate) [6] and for convenience, it is typically expressed in ampere-hours (Ah) and not in coulombs (C), despite the latter is the unit of the SI. Very often and for comparative purposes, instead of absolute capacity, electrochemists talk about gravimetric capacity, which is the capacity per gram of active material used and it is expressed in ampere-hours/gram ($Ah \cdot g^{-1}$).

Other terms of interest, most of them necessary for understanding the information compiled in the following chapters, are

- **C-rate**, or **current density**, is the rate at which a battery is discharged or charged relative to its maximum capacity (often referred to its theoretical capacity). For instance, a C-rate of 1C means that the discharge/charge current will discharge or charge the entire battery in 1 h [9].
- **Specific energy** ($Wh \cdot kg^{-1}$), also referred to as gravimetric energy density, defines the nominal energy of a battery per unit mass. It can be calculated using Eq. 1.1 and determines the battery weight required to achieve a specific electrochemical performance target [9]. At lab scale, the specific energy is often referred to the mass of the electrode's active materials only. However, in commercial systems, the weight of all inactive components (including the electrolyte, separator, current collectors and casing) is also considered and, in best case scenario, results in

specific energies three to four times lower than those based only on the weight of active materials [10].

$$Specific\ Energy\left(Wh\cdot kg^{-1}\right) = \int_{t_0}^{t} Capacity\left(Ah\cdot kg^{-1}\right)\cdot Voltage(V)\,dt$$

$$(1.1)$$

- **Specific power** (W·kg^{-1}) is the maximum available power per unit mass [9].
- **Energy density** (Wh·l^{-1}), or volumetric energy density, is the nominal energy of a battery per unit of volume. Analogously to the specific energy, it determines the battery size to achieve a specific electrochemical performance target [9].
- **Power density** (W·l^{-1}) is the maximum available power per unit of volume [9].
- **Coulombic efficiency** is the ratio of the output (charge delivered) to the input capacity (charge required) of a secondary battery [6].
- **Capacity retention** is the fraction of full capacity available from a battery under specified conditions of discharge after it has been stored for a period of time or after a number of galvanostatic cycles. A value inferior to 80% is considered as the end of the cycle life or death of the battery [6].
- **Polarization** or overpotential is the change of the potential of a cell or electrode from its equilibrium value caused by the passage of electric current [6]. In practice, it is usually calculated as the difference between the average charge and discharge voltages in a redox process.
- **Cyclic voltammetry** (CV) is a type of potentiodynamic electrochemical measurement consisting of reversibly sweeping the potential of a working electrode between two values (V$_1$–V$_2$) at a fixed scan rate (mV·s^{-1}). Each sweep (forth and back) is a cycle and these cycles may be repeated as many times as needed, leading to cyclical phases. It provides information related to the redox processes and the reversibility of the reactions involved. For a deeper understanding of electrochemical measurements, the reader is referred to other texts [11].
- **Galvanostatic cycling (or galvanostatic charge/discharge)** is a constant current measurement in which electrons flow from the working electrode to the counter electrode (and vice versa) and the resulting potential changes between the working electrode and a reference electrode are monitored over time. The current flow in one direction and the reverse direction provides the charge/discharge cycle and, similarly to the CV, these cycles may be repeated as many times as needed. Conveniently, in the field of batteries, the usual presentation of the results implies the conversion of time in specific (or gravimetric) capacity.

Batteries can be classified into two main types:

- *Primary batteries.* A primary battery is a non-rechargeable battery, that is supplied fully charged and discarded once discharged. Typical primary batteries are: *zinc–carbon (Leclanché), alkaline, silver oxide, mercury, lithium and zinc–air* [6].
- *Secondary batteries.* This class of batteries is also known as rechargeable batteries. As their name indicates, after being discharged, they can be recharged many times for reuse. Thus, conventional rechargeable batteries can bi-directionally con-

vert energy between electrical and chemical energy. There are many rechargeable batteries, which are named according to the chemistry of their electrodes or electrolytes or a combination of both, such as *lead–acid, nickel–cadmium, nickel–metal hydride, sodium–sulphur* [6], *redox flow batteries* [12], *lithium-ion batteries* [6] *and a novel group designated as beyond lithium-ion batteries* [13, 14].

In this book, we will focus on those rechargeable batteries [16, 17] which use Prussian Blue type materials at least as one of the electrodes. In particular, lithium-ion and some of the so-called beyond lithium-ion batteries using this type of systems will be revisited.

1.3 Li-ion Batteries

The appeal of lithium (Li) electrochemistry lies in the small ionic radius of lithium (0.76 Å) [15], which is beneficial for its diffusion through small tunnels in the crystal structure of the active materials that integrate the electrode, its low atomic weight (6.93 g/mol) and its negative redox potential ($E_{Li^+/Li}^+ = -3.04$ V vs NHE), which result in the high energy and power density that these batteries can store and release.

Secondary Li batteries using metallic lithium as the anode were proposed by Whittingham in the 1970s [16, 17]. Regrettably, their commercialization could never be practical as a consequence of the safety problems that arose from lithium plating at the anode surface during the charging processes, which favours dendrite growth across the electrolyte, leading to the short circuit of the cell [18, 19] and, in the worse case, its thermal runaway. In 1991, Sony solved the problem efficiently and the so-called Li-ion batteries (LIB) or 'rocking chair' technology was marketed [20]. Unlike secondary Li batteries, LIB mainly exploits the use of intercalation compounds in both electrode materials where the lithium ions 'rock' back and forth, getting inserted into the anode as the cell is charged (or forced to increase the voltage), and leaving the anode to intercalate into the cathode as the cell is spontaneously discharged upon closure of the circuit. The electrons, that travel through the external circuit, are stored in the electrode where the Li-ions are intercalated. In Fig. 1.2, a scheme of the first LIB using a layered oxide cathode ($LiMO_2$, where M = transition metal) and a graphite anode is depicted.

In general, cathode materials of LIB own their high operation voltage to a redox process in a transition metal. This voltage is determined by the metal, its oxidation state (the higher oxidation state, the higher the voltage) and by inductive effects of the anions and the crystal structure within which the transition metal accommodates [18, 19]. In addition, the kinetics of charge and discharge of the materials are related to its ionic and electronic conductivity, which mostly depend on the crystal structure and composition as well as on the sites and paths available for lithium-ion diffusion. There are excellent books [21, 22] and reviews [13, 23–25] on a plethora of materials that have been and continue to be proposed as cathodes for Li-ion batteries. We are not covering them here, but to name a few examples: $LiCoO_2$, $LiMn_2O_4$ or $LiFePO_4$

Fig. 1.2 Schematic
illustration of the first Li-ion
battery (graphite/Li$^+$
electrolyte/LiCoO$_2$) [26].
Reprinted with permission
from [26]. Copyright © 2013
American Chemical Society

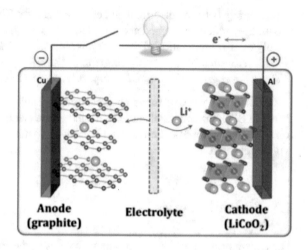

are the most common cathode materials employed in commercial batteries, where
the lithium ions are free to diffuse in 2D planes, 3D interconnected networks or 1D
tunnels, respectively.

On the other hand, anode materials used in LIB are those able to accept lithium ions
and be reduced at low voltages, in some cases even close to the lithium plating voltage.
These latter are typically non-metals such as graphite, which has been the commercial
lithium intercalation anode for 30 years, and silicon or other elements of group 4 and
5, which react by alloying with lithium and are intended to replace graphite in higher
energy density batteries [13]. Other binary and ternary compounds, for instance
oxides and sulphides, also experience reaction with lithium at low voltages. These
undergo a conversion reaction in which the lithium displaces the transition metal
and the metal gets reduced to the metal state. However, large first cycle irreversible
capacity, energy inefficiencies due to voltage hysteresis during charge and discharge
and capacity fading upon cycling [23], have mostly hindered the possible application
of these systems in commercial cells.

1.4 Beyond Li-ion Batteries

The term beyond Li-ion batteries covers a broad range of battery chemistries which
have been developed after the boom of high energy density Li-ion batteries. These
move in three different directions:

1. Using the redox properties of very light O$_2$ and S$_8$ cathodes, i.e. the so-called
   Li–air [27, 28] and Li–sulphur [29] batteries, capable of achieving much larger
   capacities than Li-ion batteries. They have promising theoretical energy densities
   approaching that of gasoline and thus are targeted towards electric vehicle (EV)
   applications. However, although a lot of understanding of their mechanism of

Table 1.1 Comparison of physical properties of Li^+, Na^+, K^+, Mg^{2+} [31], Ca^{2+} and Al^{3+} as charge carriers for rechargeable batteries

	Li^+	Na^+	K^+	Mg^{2+}	Ca^{2+}	Al^{3+}
Relative atomic mass (g/mol)	6.94	23.00	39.10	24.31	40.08	26.98
Mass-to-electron ratio	6.94	23.00	39.10	12.16	20.04	13.49
Shannon's ionic radii (Å)	0.76	1.02	1.38	0.72	1.00	0.53
E^0 versus SHE (V)	−3.04	−2.71	−2.93	−1.55	−2.87	−1.66
Theoretical capacity of metal electrodes ($mAh \cdot g^{-1}$)	3861	1166	685	2205	1337	2980

reaction has been gained, they still present many challenges related with cycle life [30] that are mainly due to the reactivity of the electrolyte and cathode electrode, the formation of insulating materials or soluble species, as well as O_2 purification to be deployed in open air.

2. Replacing Li^+ by less expensive and more abundant Na^+ or K^+ ions, i.e. switching to Na- and K-ion batteries, despite the decrease in energy density that this implies. This lower energy density derives from the intercalation of heavier ions, often using bulkier hosts, and the not so low reduction potential of the metal, which in the case of Na^+ prevents the use of very low insertion voltage anodes such as graphite or silicon and requires the search of alternative materials [32]. Nevertheless, these technologies benefit from the knowledge acquired for Li-ion batteries, given the similarity between the chemistry of Li, Na and K. Regarding the applications, the interest of Na- and K-ion batteries is mostly focused on grid-scale applications, where the weight and volume are not a constraint.

3. Replacing monovalent Li^+ by more abundant divalent or trivalent metals such as Mg^{2+} [33, 34], Ca^{2+} [35] or Al^{3+} [36] to decrease the cost and increase significantly the energy density in case that metallic anodes are used. Unlike Na- and K-ion batteries, divalent and trivalent metals are not expected to form dendrites. Nonetheless, avoiding the current collector corrosion and finding a suitable electrolyte are their major challenges. These technologies are at early stages of research, so big improvements are still expected from them to become a reality.

In Table 1.1, the physical properties (relative atomic weight, mass-to-electron ratio, Shannon's ionic radii, standard potential and the theoretical capacity) of different alkali and alkali-earth metals are listed.

With this, we conclude the first chapter. In the following chapters, given the structure and redox properties of Prussian Blue (PB) and Prussian Blue Analogues (PBA) (Chap. 2), we will witness how they have mostly been investigated as cathode materials in Li-, Na- and K-ion batteries (Chaps. 3—aqueous batteries—and 4—non-aqueous batteries—). Interestingly, some recent studies have opened the doors to their evaluation also as anodes, discovering that they do not follow the conventional intercalation mechanism (Chap. 4). Like many other materials, PB and its derivatives present several drawbacks. Nevertheless, their simplicity of synthesis along with their competitive performance as cathodes is pushing them towards commercialization.

References

1. N. Kipnis, Ann. Sci. **44**(2), 107–142 (1987)
2. J. Newman, K.E. Thomas-Alyea, *Electrochemical Systems*, 3rd edn. (Wiley, Hoboken, 2004)
3. A. Frood, *Riddle of Baghdad Battery* (BBC News, 27 Feb 2003), http://news.bbc.co.uk/2/hi/science/nature/2804257.stm
4. A.J. McEvoy, *EPJ Web of Conferences*, vol. 54 (2013), p. 01018
5. G.G. Botte, *Batteries: Basic Principles, Technologies and Modeling. Encyclopedia of Electrochemistry* (Wiley, 2007)
6. D. Linden, T.B. Reddy, *Handbook of Batteries*, 3rd edn. (McGraw Hill Handbooks, New York, 1995)
7. J.B. Goodenough, Acc. Chem. Res. **46**(5), 1053–1061 (2013)
8. B. Dunn, H. Kamath, J.-M. Tarascon, Science **334**, 928 (2011)
9. A Guide to Understanding Battery Specifications. MIT Electric Vehicle Team (2008), http://web.mit.edu/evt/summary_battery_specifications.pdf
10. Y. Gogotsi, P. Simon, Science **334**, 917–918 (2011)
11. A.J. Bard, L.R. Faulkner, *Electrochemical Methods: Fundamentals and Applications* (Wiley, New York, 2000)
12. X. Luo, J. Wang, M. Dooner, J. Clarke. Appl. Energy **137**, 511–536 (2015)
13. J.W. Choi, D. Aurbach, Nat. Rev. Mat. (2016). https://doi.org/10.1038/natrevmats.2016.13
14. M. Armand, J.-M. Tarascon, Nature **451**, 652–657 (2008)
15. R.D. Shannon, Acta Cryst. **A32**, 751 (1976)
16. M.S. Whittingham, Science **192**, 1226 (1976)
17. M.S. Whittingham, Chalcogenide battery, US Patent 4009052
18. J.B. Goodenough, Y. Kim, Chem. Mater. **22**(3), 587–603 (2010)
19. V. Srinivasan, *AIP Conference Proceedings* (2008)
20. T. Nagura, K. Tozawa, Prog. Batter. Sol. Cells **9**, 209 (1990)
21. G.A. Nazri, P. Gianfranco (eds.), *Lithium Batteries: Science and Technology* (Kluwer Academic Publishers, Norwell, 2003)
22. M. Yoshio, R.J. Brodd, A. Kozawa (eds.), *Li-ion Batteries: Science and Technologies* (Springer, New York, 2009)
23. N. Nitta, F. Wu, J.T. Lee, G. Yushin, Mat. Today **18**(5), 252–264 (2015)
24. G. Jeong, Y.-U. Kim, H. Kim, Y.-J. Kim, H.-J. Sohn, Energy Environ. Sci. **4**, 1986 (2011)
25. J.M. Tarascon, M. Armand, Nature **414**, 359–367 (2001)
26. J.B. Goodenough, K.-S. Park, J. Am. Chem. Soc. **135**(4), 1167–1176 (2013)
27. K.M. Abraham, Z. Jiang, J. Electrochem. Soc. **143**(1), 1–5 (1996)
28. T. Liu, M. Leskes, W. Yu, A.J. Moore, L. Zhou, P.M. Bayley, G. Kim, C.P. Grey, Science **350**(6260), 530–533 (2015)
29. P. Bruce, S.A. Freunberger, L.J. Hardwick, J.-M. Tarascon, Nat. Mater. **11**, 19–29 (2012)
30. D. Aurbach, B.D. McCloskey, L.F. Nazar, P.G. Bruce, Nat. Energy **1**, 1–11 (2016)
31. N. Yabuuchi, K. Kubota, M. Dahbi, S. Komaba, Chem. Rev. **114**(23), 11636–11682 (2014)
32. M.Á. Muñoz-Márquez, D. Saurel, J.L. Gómez-Cámer, M. Casas-Cabanas, E. Castillo-Martínez, T. Rojo, Adv. Energy Mater. **7**(20), 1700034 (2017)
33. P. Novak, W. Scheifele, O. Haas, J. Power Sources **54**, 479–482 (1995)
34. D. Aurbach, G.S. Suresh, E. Levi, A. Mitelman, O. Mizrahi, O. Chusid, M. Brunelli, Adv. Mater. **19**, 4260–4267 (2007)
35. A. Ponrouch, C. Frontera, F. Bardé, M.R. Palacín, Nat. Mater. **15**, 169–172 (2016)
36. N. Jayaprakash, S.K. Das, L.A. Archer, Chem. Commun. **47**, 12610–12612 (2011)

Chapter 2
Prussian Blue and Its Analogues. Structure, Characterization and Applications

2.1 The Origins of Prussian Blue

Prussian Blue is a mixed-valence polynuclear transition metal cyanide complex [1, 2] and it can be considered the first synthetic coordination compound [3]. Its discovery dates back to early eighteenth century, around 1704, and its invention has been attributed to a colour manufacturer from Berlin named Diesbach, who apparently obtained the blue compound accidentally when he was preparing another pigment [4].

Although the first reference on Prussian Blue was anonymously reported on 1710 [5], the method of preparation remained secret until 1724 [6] as a result of its profitable use as a pigment for painters. In fact, Prussian Blue is recognized as the first synthetic pigment. The original method reported by Woodward [6] for Prussian Blue preparation consisted of the addition of a blend of crude alum ($AlK(SO_4)_2 \cdot 12H_2O$) and green vitriol ($FeSO_4 \cdot xH_2O$) to an alkali solution (essentially K_2CO_3) previously calcinated with cattle blood (where CN^- groups are present). The acidification of the resulting greenish precipitate with HCl yielded the deep blue compound, namely Prussian Blue [7].

Over the years and with a better comprehension of the subjacent chemical reactions occurring, thanks to the effort of several scientists [8], the procedure was greatly improved and the methods of synthesis were standardized, as it is detailed in the subsequent section.

2.2 The Formula and Structure of Prussian Blue

Despite the blue pigment had been known since 1704, the first structural determination of the complex was not achieved until 1936 [9]. According to the powder X-ray pattern analysis, Keggin and Miles could satisfactorily describe the structure

© The Author(s) 2018
M. J. Piernas Muñoz and E. Castillo Martínez, *Prussian Blue Based Batteries*,
SpringerBriefs in Applied Sciences and Technology,
https://doi.org/10.1007/978-3-319-91488-6_2

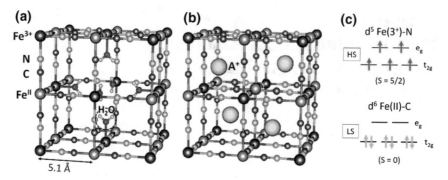

Fig. 2.1 Structural forms of Prussian blue: **a** '*Insoluble*' Prussian Blue, $Fe_4[Fe(CN)_6]_3 \cdot xH_2O$ and **b** '*Soluble*' Prussian Blue, $AFe^{3+}[Fe^{II}(CN)_6] \cdot xH_2O$, where A^+ occupies zeolitic-type sites. Reproduced and adapted with permission from Ref. [10]. Copyright © 2017, Royal Society of Chemistry. **c** Electronic configuration of LS Fe^{II}–C and HS Fe^{3+}–N in PB

of Prussian Blue (PB) as a 3D cubic network composed by iron atoms (ferrous (Fe^{II}) and ferric (Fe^{3+})) alternatively located at the corners of a small cube of 5.1 Å edge and linked by bidentate –C≡N– ligands, as Fig. 2.1a, b illustrate. In general, the metalorganic framework is mainly constructed by Fe^{II}–C≡N–Fe^{3+} units all along the three directions of the space, featuring zeolitic sites capable to host species with an ionic radio of 1.6 Å [11]. Each crystallographic unit cell comprises eight cubes and has a lattice parameter of about 10.2 Å. Interestingly, the characteristic intense blue colour of Prussian Blue is due to the mixed-valence nature of the compound (containing Fe^{II} and Fe^{3+}, as Fig. 2.1c illustrates) and originates in an electronic transition from the Fe^{II} to the Fe^{3+} ion caused by the absorption of orange-red light of the visible spectrum (680–730 nm) [12]. The reasons for using different nomenclature to refer to the different types of iron (Fe^{II} and Fe^{3+}) will be discussed later.

Depending on the synthesis methodology and the reactants used, alkali metal ions and a variable amount of water may occupy the zeolitic sites, yielding slightly different varieties of the compound that include '*insoluble*' and '*soluble*' Prussian Blue and what was known as Turnbull's Blue [13].

2.2.1 '*Insoluble*' and '*Soluble*' Prussian Blue

The classical methodology to obtain Prussian Blue is by *co-precipitation*. In a single-step reaction, an aqueous solution of a hexacyanoferrate (II) salt,[1] $A_4[Fe^{II}(CN)_6]$ (where A = alkali metal, typically K), is mixed with an aqueous solution containing an iron (+3) salt, such as ferric chloride, $FeCl_3$, leading to the formation of a deep blue precipitate [14]. Notwithstanding the simplicity of the synthesis, the extremely

---

[1]Another accepted nomenclature for the hexacyanoferrate (II) fragment, $[Fe(CN)_6]$, is "ferrocyanide".

small particle size makes, however, particularly difficult the isolation of the product from impurities and adsorbed water.

'*Insoluble*' **Prussian Blue** is formed as long as there is an excess of iron (+3) salt present in the medium of reaction. It contains very little (if any) alkali metal in the zeolitic sites, but it accommodates relatively large amounts of water [13]. The formula for this *alkali metal-free* complex has been established as iron (III) hexacyanoferrate (II), i.e. $Fe_4^{3+}\left[Fe^{II}(CN)_6\right]_3 \cdot xH_2O$ (where "x" is variable, although usually oscillates in the range of 14–16), and its preparation can be summarized according to Eq. 2.1[2]:

$$4\,Fe^{3+}Cl_3\,(aq) + 3\,A_4\left[Fe^{II}(CN)_6\right]\,(aq)\ \rightarrow\ Fe_4^{3+}\left[Fe^{II}(CN)_6\right]_3 \cdot xH_2O(s) + 12\,ACl\,(aq)$$

$$(2.1)$$

Based on data extracted from ^{57}Fe-Mössbauer experiments [15] and diffraction techniques [16, 17], it was established that Fe^{II} atoms in the cubic framework of *insoluble* Prussian Blue are in a low spin (LS) d^6 electronic configuration and covalently coordinated to six carbon atoms of the –CN ligands (see Fig. 2.1c bottom), while Fe^{3+} centres are high spin (HS) d^5 and are 'ionically' bonded to the nitrogen atoms (Fig. 2.1c top). Hence their different nomenclature. Indeed, the strong covalent bond between Fe^{II} and the six cyanide ligands is reflected in the fact that $[Fe(CN)_6]^{4-}$ units remain stable in aqueous solution until very acidic pH, as well as by the existence of $[Fe(CN)_6]^{4-}$ vacancies in the solid. In *insoluble* Prussian Blue, a quarter of the $[Fe^{II}(CN)_6]^{4-}$ sites are vacant.[3] As a result, Fe^{3+} cations are also coordinated to water molecules through oxygen, occupying these latter the N-vacant sites (24e sites). Therefore, Fe^{3+} is on average coordinated by 4.5 nitrogen atoms and 1.5 oxygen atoms [16, 17] (see Fig. 2.1a for a better understanding). As an example, a unit cell of $Fe_4^{3+}\left[Fe^{II}(CN)_6\right]_3 \cdot 14H_2O$ would contain 18 carbon atoms bonded to $3Fe^{II}$, 18 nitrogen atoms and 6 oxygen atoms (H_2O) coordinated to $4Fe^{3+}$ building the cubic framework, and 8 non-coordinated water molecules occupying zeolitic-type positions [18] within the lattice structure.

Conversely, if the reagents are mixed stoichiometrically or with an excess of hexacyanoferrate (II), the so-called '*soluble*' **Prussian Blue** is obtained. Unlike the insoluble species, this compound accommodates alkali metal ions in its structure [13]. The formula for this *alkali-containing* Prussian Blue can be expressed as $AFe^{3+}[Fe^{II}(CN)_6] \cdot xH_2O$ (where A = alkali metal, typically K; "x" is variable) and its synthesis reaction can be described by Eq. 2.2:

$$Fe^{3+}Cl_3\,(aq) + A_4\left[Fe^{II}(CN)_6\right]\,(aq)\ \rightarrow\ AFe^{3+}\left[Fe^{II}(CN)_6\right] \cdot xH_2O(s) + 3\,ACl\,(aq)$$

$$(2.2)$$

Despite the term coined to refer to this species, the material is not really soluble (as Eq. 2.2 shows). Its name is due to the fact that in aqueous solutions, $AFe^{3+}[Fe^{II}(CN)_6] \cdot xH_2O$ can be easily dispersed (peptized) given the colloidal nature of the particles [19], generating a blue suspension that resembles a solution.

---

[2]Please, note that the oxidation states of the irons are indicated in all the equations of the present chapter for further clarification.

[3]$[Fe(CN)_6]^{4-}$ vacancies are commonly represented as □.

Although the lattice is quite similar to that of the *insoluble* compound (LS Fe^{II}–C–N–HS Fe^{3+}), there are slight variations between them. It is generally accepted that there are no $[Fe^{II}(CN)_6]^{4-}$ vacancies in the structure of *soluble* Prussian Blue (as Fig. 2.1b depicts), and thus only non-coordinated water, which occupies approximately half of the zeolitic-type sites can be present. The remaining zeolitic positions, ideally the other half, host alkali ions A^+ (A = alkali metal, typically potassium) [1, 2, 9, 13, 18] that act as counter-anions to compensate the negative charge of $\{Fe^{3+}[Fe^{II}(CN)_6]\}^-$ and so maintain electrical neutrality [18]. Nevertheless, a synchrotron X-ray diffraction study from 2008 showed that 25% of the $[Fe^{II}(CN)_6]^{4-}$ sites are vacant in the *soluble* species (as it occurs in the "insoluble" species) [20], at the same time that K^+ ions occupy 24e sites close to those of coordinated crystalline water. Based on that, the authors of that work proposed a new stoichiometry $Fe_4^{3+}[Fe^{II}(CN)_6]_3 \cdot [K_h^+ \cdot OH_h^-] \cdot x H_2O$ to describe the compound and strongly recommended to reconsider the terms 'soluble' and 'insoluble', as they may be inappropriate to refer to the same material (from a structural point of view), the only difference being the presence or not of K^+ ions.

Regardless of the amount of interstitial water and $[Fe(CN)_6]^{4-}$ vacancies, Prussian Blue crystallizes in the cubic space group Fm-$3m$. On it, Fe^{II} from $[Fe(CN)_6]^{4-}$ units and Fe^{3+} are allocated in 4a and 4b Wyckoff positions, respectively. Carbon (C), nitrogen (N) and the oxygen of coordinated water accommodate in 24e positions; whereas K ions (or any replacing cation) as well as interstitial water stay in the zeolitic cavities, typically in the centre 8c or slightly offset at 32f position, or in 24d sites (here, only cations, not water) [17, 20, 21]. See Fig. 4.11b (Chap. 4), for further clarification.

Even though the material is old, it is evident that there is still some controversy regarding its precise structure and composition, which depends to a large extent on the synthesis method. However, the classical models and nomenclatures, i.e. 'insoluble'/'soluble', are still widely accepted and used by the scientific community and will also be adopted in this book. In any case, the majority of researchers that use Prussian Blue and related materials perform elemental analyses and atomic absorption/emission spectroscopy or inductively coupled plasma (ICP) measurements to precisely determine the number of $[Fe^{II}(CN)_6]^{4-}$ vacancies and the alkali cations per transition metal for each specific material. As we will see in Chaps. 3 and 4, most of the Prussian Blue materials reported for battery research have intermediate compositions between those of the ideal soluble (vacancy free) and ideal insoluble (1/4 of $Fe^{II}(CN)_6]^{4-}$ vacancies and no alkali cation) formulations.

2.2.2 Turnbull's Blue

Alternatively to Prussian Blue, the addition of a solution containing an iron (+2) salt (for example iron dichloride, $FeCl_2$) to a solution of hexacyanoferrate (III),[4]

---

[4]Another accepted terminology for hexacyanoferrate (III) is ferricyanide. For example, $K_3[Fe(CN)_6]$ is referred as potassium ferricyanide.

$A_3[Fe^{III}(CN)_6]$ (A = alkali ion, typically K), was supposed to produce a salt called Turnbull's Blue [13] (see Eqs. 2.3 and 2.4):

$$Fe^{2+}Cl_2 + A_3\left[Fe^{III}(CN)_6\right] \rightarrow AFe^{2+}\left[Fe^{III}(CN)_6\right]\cdot xH_2O + 2\,ACl \quad \text{(soluble)} \quad (2.3)$$

$$3Fe^{2+}(aq) + 2\left[Fe^{III}(CN)_6\right]^{3-}(aq) \rightarrow Fe_3^{2+}\left[Fe^{III}(CN)_6\right]_2\cdot xH_2O\,(s) \quad \text{(insoluble)} \tag{2.4}$$

Although originally it was assumed that the Turnbull's blue was an iron (II) hexacyanoferrate (III), either $AFe^{2+}[Fe^{III}(CN)_6]\cdot xH_2O$ or $Fe_3^{2+}\left[Fe^{III}(CN)_6\right]_2\cdot xH_2O$, X-ray diffraction [22] and Mössbauer [15, 23] spectroscopy studies have unequivocally demonstrated that this reaction produces iron (III) hexacyanoferrate (II) as product, i.e. Prussian blue ($AFe^{3+}[Fe^{II}(CN)_6]\cdot xH_2O$ or $Fe_4^{3+}\left[Fe^{II}(CN)_6\right]_3\cdot xH_2O$). A phenomenon that is attributed to a charge transfer from HS Fe^{2+} to LS Fe^{III} or to a 180° flipping of the CN ligand at the instant of the combination of the precursors [15].

Having explained the differences and similarities existent among the varieties of Prussian Blue and Turnbull's Blue, we will focus now on those species related to 'soluble' Prussian Blue.

2.3 The Reduced and Oxidized Forms of Prussian Blue

There are two more species or phases closely related to 'soluble' Prussian Blue: the so-called *Berlin Green* and *Prussian White*.

Berlin Green, sometimes also known as Prussian Green, is typically formulated as $Fe^{3+}[Fe^{III}(CN)_6]$, despite there is some residual potassium in its structure, as detailed later. The arrangement of the iron, carbon and nitrogen atoms in this phase remains the same than in Prussian Blue. Though it differs from the former in the oxidation state of iron, since all iron atoms are in principle in oxidation state +3, and because it does not allocate alkali ions [24] (see Fig. 2.2, Berlin Green). In a simplified manner, Berlin Green can be defined as the totally oxidized phase of Prussian Blue. It can be obtained by direct synthesis, simply mixing the alkali ferricyanide with an Fe^{3+} salt (such as $FeCl_3$) [13], as Eq. 2.5 illustrates:

$$A_3\left[Fe^{III}(CN)_6\right] + Fe^{3+}Cl_3 \leftrightarrow Fe^{3+}\left[Fe^{III}(CN)_6\right](aq) + 3\,ACl\,(aq) \tag{2.5}$$

Prussian White, also referred to as Everitt's salt, is ideally $K_2Fe^{2+}[Fe^{II}(CN)_6]$ (A = alkali ion, typically K^+). The Fe–C–N–Fe framework of Prussian Blue is maintained too in this species albeit all the iron atoms are in oxidation state +2 and the zeolitic-type sites are fully occupied by alkali ions [13, 24] (see Fig. 2.2, Prussian White). Analogously to Berlin Green, Prussian White can be considered the reduced phase of Prussian Blue. It can be formed by precipitation upon mixing an aqueous

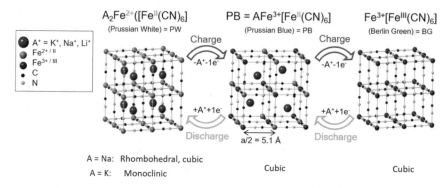

Fig. 2.2 Prussian Blue structure and its corresponding oxidized (Berlin Green) and reduced (Prussian White) species

solution of hexacyanoferrate(II) with an iron (+2) salt solution [25], for example, $FeCl_2$, as shown in Eq. 2.6

$$A_4\left[Fe^{II}(CN)_6\right]\cdot x\,H_2O + Fe^{2+}Cl_2 \leftrightarrow A_2Fe^{2+}\left[Fe^{II}(CN)_6\right](aq) + 2\,ACl \qquad (2.6)$$

Another interesting approach to prepare Prussian White and Berlin Green is by electrochemical reduction or oxidation of Prussian Blue [26], what involves cationic and electronic transport within the structure, as the red and green arrows indicate in Fig. 2.2. These electrochemical changes will be addressed more in-depth in the subsequent chapters (Chaps. 3 and 4).

Only via electrochemistry, another phase called Prussian Yellow can be obtained. Early electrochemical studies showed that Prussian Blue thin films became either green [26] or brown [27], respectively, upon oxidation at 1.0 or 1.4 V versus SCE (saturated calomel electrode). Because the brown film (assigned to Prussian Yellow) is not stable and is readily reduced by water, the most commonly known oxidized phase is the green phase or Berlin Green. In other words, *Prussian Yellow* is really the 'totally' oxidized phase of Prussian Blue, $Fe^{3+}[Fe^{III}(CN)_6]$. The Fe–C–N–Fe Prussian Blue framework is also retained in this compound, but all the irons atoms are in oxidation state +3. On the other hand, *Berlin Green* is actually $K_{1/3}Fe^{3+}[Fe^{II}(CN)_6]_{1/3}[Fe^{III}(CN)_6]_{2/3}$. It differs from the previous in the oxidation state of iron, being most of them in oxidation state +3 but some are +2. Because of that, it still allocates one-third of alkali ions [24]. Nonetheless, given the lower stability of Prussian Yellow and the lack of precise composition determination, from now on, and regardless of being either the partially or totally oxidized form, we will refer to the oxidized phase as Berlin Green (BG), unless the authors of these works, specifically call it and/or show it to be Prussian Yellow.

Along with the changes in oxidation state among the Prussian Blue related phases, structural changes often occur. Prussian Blue generally adopts a cubic face-centred structure (space group *Fm-3m*) with lattice parameter $a =$ ca. 10.178 Å, giving a

Fe^{3+}–N≡C–Fe^{II} bond length of about 5.1 Å. Berlin Green preserves the cubic symmetry [28]. Prussian White, however, can crystallize in cubic structure [29] or in lower symmetry, either monoclinic [30] or rhombohedral [31, 32], depending on the alkali content resultant from the specific synthesis method employed. The monoclinic phase of potassium Prussian White has been obtained from a single iron source, by dissolving $K_4Fe(CN)_6 \cdot 3H_2O$ in deaerated distilled water, and subsequently subjecting the mixture to 160 °C for 48 h [30]. When the sodiated reactant, $Na_4Fe(CN)_6 \cdot 10H_2O$, is used instead in presence of ascorbic acid and the solution is maintained at 140 °C for 20 h, the rhombohedral phase of sodium Prussian White is achieved [31]. On the other hand, cubic phases can be prepared in the presence of ascorbic acid and N_2 atmosphere, either by reacting $Na_4Fe(CN)_6 \cdot 10H_2O$ with HCl 1 M at 80 °C for 4 h [32] or with $FeCl_2$ at 60 °C for 4 h [29].

2.4   Electronic, Vibrational and Nuclear Properties of Prussian Blue and Related Phases

As mentioned previously, the intense deep blue colour of Prussian Blue arises from the charge transfer transition in the Fe^{II}–C–N–Fe^{3+} unit when red light of the visible spectrum is absorbed, as a result of the displacement of an electron from the $(t_{2g})^6$ orbitals of Fe^{II} to the $(t_{2g})^5$ orbitals of Fe^{3+} (see Eq. 2.7):

$$Fe^{II} - C + Fe^{3+} - N \; \rightarrow \; Fe^{III} - C + Fe^{2+} - N \qquad (2.7)$$

Any change in the electronic configuration, thus, implies the shift or loss of colour. This explains the green colour observed for Berlin Green, which is due to electronic transitions, and the lack of colour in the Prussian White phase, since it does not experience crystal-field transitions in the visible region [24].

Aside from visually and, in the case of comparing phases with different symmetry, by X-ray diffraction (XRD), other helpful techniques to differentiate among Prussian White, Prussian Blue and Berlin Green are those related to molecular vibrations, i.e. infrared spectroscopy (IR) and its complementary, Raman spectroscopy. Given the sensitivity of the cyanide stretching frequency (ν(–C≡N)) to the bonding and environment, it can be used as a fingerprint to track and compare the different Prussian Blue species (see Table 2.1). In IR, the characteristic ν(–C≡N) band for potassium Prussian Blue has been reported at ca. 2080 cm^{-1} [33]. Instead, deviations towards lower wavenumbers (2067 cm^{-1}) [29] and higher wavenumbers (2090 cm^{-1}) [34] have been respectively observed for potassium Prussian White and Berlin Green, suggesting longer C≡N distances when the oxidation state for both iron atoms is close to +2 and shorter C≡N distances when both get oxidized to +3. As for the Raman, two C≡N bands are distinguished for potassium Prussian Blue at 2152 and 2091 cm^{-1}, while three bands are detected for Prussian White (2128, 2093 and 2041 cm^{-1}) and Berlin Green (2149, 2131 and 2091 cm^{-1}) [35]. Like in IR

Table 2.1 Typical IR frequencies and Raman shifts of potassium Prussian White, potassium Prussian Blue and Berlin Green

	Prussian White	Prussian Blue	Berlin Green
IR frequencies (cm^{-1})	2067	2080	2090
Raman shifts (cm^{-1})	2128, 2093, 2041	2152, 2091	2149, 2131, 2091

Table 2.2 Typical ^{57}Fe-Mössbauer parameters δ and Δ for potassium Prussian White, potassium Prussian Blue and Berlin Green (δ =isomer shift relative to Fe bcc, Δ =quadrupolar splitting). Note that "LS" refers to low spin and "HS" to high spin

Compound	Assignation	Isomer shift, δ (mm·s^{-1})	Quadrupolar, Δ (mm·s^{-1})
$K_2Fe^{2+}[Fe^{II}(CN)_6]$ [30] (Prussian White)	LS Fe^{II}	−0.06(2)	0
	HS Fe^{2+}	1.16(1)	1.61(3)
$KFe^{3+}[Fe^{II}(CN)_6]$ [35] (Prussian Blue)	LS Fe^{II}	−0.140(1)	0.042(1)
	HS Fe^{3+}	0.400(1)	0.000(5)
$Fe^{3+}[Fe^{III}(CN)_6]$ [35] (Berlin Green)	LS Fe^{III}	−0.168(1)	0.119(3)
	HS Fe^{3+}	0.381(1)	0.448(2)

spectroscopy, in general, red-shift is observed when the material gets reduced and the opposite trend (blue-shift) upon oxidation.

Another common technique and probably the most effective to follow the evolution of the iron oxidation state in Prussian Blue and Prussian Blue related materials is ^{57}Fe-Mössbauer spectroscopy. As displayed in Table 2.2, the Mössbauer parameters significantly vary from one type of iron to another in each of the species, allowing as well to easily differentiate Prussian blue from Prussian White or Berlin Green. Indeed, Mossbauer spectroscopy has enabled to establish that in the three phases (BG, PB and PW), the iron bonded to C is in LS configuration, regardless of being Fe^{II} (d^6) or Fe^{III} (d^5); while the iron bonded to the N is in HS configuration, independently of being Fe^{2+} (d^6) or Fe^{3+} (d^5).

Similarly, though to a lesser extent (probably as a result of its surface analysis nature), XPS (X-ray photoelectron spectroscopy) has also been utilized for tracing the oxidation state of iron in this type of compounds [29]. However, such details will not be discussed here.

2.5 Prussian Blue Analogues. Other Transition Metal Hexacyanides

The above detailed purely iron-based Prussian Blue and related compounds are not the only transition-metal cyanide complexes known. Since the discovery of the material, an extensive variety of compounds have been synthesized by substitution of iron by other transition metals, often at the N-coordinated site and more rarely at the C-coordinated sites (due to the high stability of the $[Fe(CN)_6]^{4-}$ unit), as well as by playing with the counter-anions. This has created the family of complexes denominated as Prussian Blue Analogues (PBA) [36, 37], whose general formula can be described as $AM[M'(CN)_6] \cdot xH_2O$ (being usually A = Li, Na, K; M, M' = transition metal; sometimes M = M'; typically M' = Fe). Among others, some the most frequently employed metals in PBA are Cr, Mn, Fe, Co, Ni and Cu. It is important to note that PBA also receive the name of hexacyanoferrates (abbreviated as HCF) if M' = Fe, or hexacyanometallates (abbreviated as HCM) if M' = other transition metal. For short, these acronyms, HCF and HCM will be predominantly used in subsequent chapters to refer to them.

PBA usually adopt the cubic structure where M'^{II} and M^{3+} (or M'^{III} and M^{2+} depending on the metals) coexist on alternate corners of a small cube connected by linear cyanide anions along the three directions of space, giving a M^{3+}–N≡C–M'^{II} or M^{2+}–N≡C–M'^{III} bond length close to 5 Å, and whose large empty cavities would be half-occupied by alkali ions [37]. The most stable electronic configuration for each metal-ligand pair will determine whether they are HS bonded to N or LS linked to C when both are possible, with even linkage isomerization (180° ligand flipping) occurring in some cases [38]. Interestingly, not all the PBA adopt the structure above described. For instance, zinc-hexacyanoferrate (Zn-HCF) presents a slightly different structure (see Fig. 2.3a). $Na_2Zn_3[Fe^{III}(CN)_6]_2$ crystallizes in a rhombohedral lattice where Fe ions are bridged by six cyanides in an octahedral coordination, like in PB, whereas Zn^{2+} occupies tetrahedral sites and is coordinated by the nitrogen atoms [39]. In this structure, K^+ and H_2O are also accommodated in the zeolitic cavities, although these are not cubic. The cavities have diameters of 5.1 Å along the c-axis and they are much larger in the ab-plane with diameters in between 8.3 and 12.7 Å.

Just like Prussian blue related species, PBA can be investigated and differentiated by all the techniques above mentioned (XRD, IR, Raman, Mössbauer, XPS) and some others, such as XAS (X-ray absorption spectroscopy). In contrast to Mössbauer spectroscopy, which is limited to very few elements having an isotope suitable and susceptible to experience the Mössbauer effect (such as Fe), XPS and XAS can be applied to analyze a wider number of transition metals. Especially in the case of XAS, some more examples will be discussed in Chapter 4 in addition to the study described below [40].

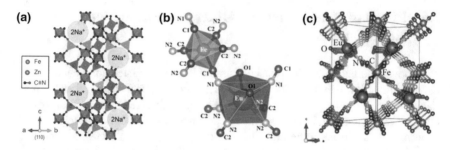

Fig. 2.3 a Unit cell structure of NaZn-HCF. Reproduced from Ref. [41]. Copyright © 2012, Royal Society of Chemistry. **b** Local coordination structure in Eu[Fe(CN)$_6$]·4H$_2$O composed of hexacyaometallate ferricyanide and square antiprism Eu, which are bridged with two crystallographically independent cyanide moieties (C1N1 and C2N2), **c** crystal structures of Eu[Fe(CN)$_6$]·4H$_2$O along the [1 1 0] direction. Solid lines in the image indicate the unit cell Reprinted with permission from Ref. [43]. Copyright © 2014, American Chemical Society

For solid solutions of the type Na$_x$(M$_h$, M$_g$)[Fe(CN)$_6$]$_y$ ($1.40 < x < 1.60$ and $0.85 < y < 0.90$), where a host metal (M$_h$ = Co, Ni, Mn) is doped with ca. 20–25% of a guest metal (M$_g$ = Mn, Co, Ni; respectively), XAS analyses have revealed that the local structure of the guest metal is essentially the same as in a non-doped compound Na$_x$M$_g$[Fe(CN)$_6$]$_y$, with hardly any change in the M$_g$–C or M$_g$–N distances [40]. This invariability on the local structure of the guest is mostly attributed to (i) the very stable electronic configuration of [Fe(CN)$_6$]$^{4-}$, which is almost unaltered by the nature of the other metal M and (ii) the flexibility of the network, which also contains a considerable number of [Fe(CN)$_6$]$^{4-}$ vacancies and is able to accommodate changes in the M–N distance by axial displacement of the M$_g$ (involvig off-axial rotation of the [Fe(CN)$_6$]$^{4-}$ units). Curiously, all this contrasts significantly with what has been observed for other type of materials, such as layered oxides (whose M$_g$–O distances change), thus evidencing the structural stability of PBA.

Similarly to PBA, Prussian White Analogue (M^{2+}–N≡C–M$'^{II}$) and Berlin Green Analogue (M^{3+}–N≡C–M$'^{III}$) phases can also be prepared. For instance, solid solutions of two different transition metals in M, such as K$_2$Fe$_{1-x}$Ni$_x$[Fe(CN)$_6$] have been synthesized [42]. Another unusual example is the EuFe(CN)$_6$·4H$_2$O, a BGA that does not adopt a PBA-type structure but an orthorhombic structure (*Cmcm*), where Eu is coordinated by the 6 N of –CN– and by 2 O from water molecules [43], as Fig. 2.3b, c depict.

In any case, all the possible combinations of metals and cations (alkali, alkaliearth and others) generate a wide number of PWA/PBA/BGA complexes that can be explored in diverse applications, as we will see in the following section and subsequent chapters.

Table 2.3 Summary table of the main applications of Prussian Blue materials and their analogues

Prussian Blue applications

		Examples
Analytical	Sensors	Tl^+, Cs^+, K^+, NH_4^+, Rb^+, humidity, cyt.[a] C, etc.
	Molecular sieves	Gas adsorption (H_2O, methanol, …), radioactive decontamination (^{137}Cs, ^{201}Tl), water desalination [44], etc.
Magnetic	Magnetic properties	Magnetic ordering, magnets
Electrochemical	Electrochromic devices	Electrochromic windows
	Catalyzers	Oxygen reduction reaction
	Electrochemical energy storage	Batteries

[a]cyt. = cytochrome

2.6 Applications of Prussian Blue and Its Analogues

The simplicity of synthesis, which is commonly carried out in aqueous media and at room temperature, and the abundance of the elements that generally compose them, make Prussian Blue and their analogues attractive materials for commercial applications. On top of that, the 3D metalorganic framework structure with open channels that these family of compound possess result in an exceptional variety of physical and chemical properties, what has fostered an intense research towards different applications, of which several examples are collected in Table 2.3.

Some of the applications of Prussian Blue and analogues exploit their analytical features. Because certain species can selectively intercalate into the channels or zeolitic-type sites of Prussian Blue type structures, these compounds have been proposed as potential sensors and molecular sieves.

Hexacyanoferrate-based sensors were first developed for thallium (Tl^+), caesium (Cs^+) and potassium (K^+). The number of cationic analytes was soon broadened, incorporating among others ammonium (NH_4^+), rubidium (Rb^+) and other mono- and divalent cations. Electrochemical techniques, either potentiometry or amperometry, were used in most of the cases for the analytical determination [45]. Furthermore, it has been found that Prussian Blue can be used as a sensor for easily oxidizable compounds, including water vapour or humidity, methanol or dichloroethane vapours, cytochrome C, persulphate, hydrazine and catecholamine [46].

During the last century, the adsorption properties of several hexacyanometallates have also been explored. By removing the water molecules accommodated in the zeolitic-type sites of the structural lattice, these materials have proved their capacity to adsorb different molecules that include solvents such as water, methanol or ethanol [47, 48] and gases, as CO, H_2 [49], N_2 and C_2H_4 [50].

Besides, the high affinity showed by Prussian Blue for cesium and thallium cations has been very effective in the medical treatment of people who had become internally contaminated with radioactive caesium (^{137}Cs) and thallium (^{201}Tl) [51]. By oral administration of Prussian Blue (Radiogardase®) [52], radioactive caesium or thallium cations are trapped by ionic exchange with potassium.

The magnetic properties of this family of compounds have also been extensively investigated. The magnetic susceptibility of Prussian Blue revealed its paramagnetism at room temperature [53], and the subsequent magnetization measurements at low temperatures exhibited a ferromagnetic behaviour with a Curie temperature T_C = 5.5 K [15], being both phenomena attributed to the presence of Fe^{3+}. By appropriately combining the transition metals, hexacyanometallates with magnetic ordering temperatures as high as room temperature and above can be prepared [54]. Moreover, there is a growing interest in hexacyanometallates applied to the development of functionalized magnets, whose magnetic properties can be controlled by external stimuli [55].

In addition to the previous applications, Prussian Blue and its analogues have proved to exhibit electrochemical activity.

In 1978, Neff pioneered the electrochemical deposition of thin films of Prussian Blue, describing as well the reversible colour change observed in the material when it was electrochemically oxidized or reduced, as will be detailed in the next chapter [56]. Since the 80s, the electrochromism (or colour change upon electrochemical modulation) of Prussian Blue films has gained significance for its potential use in smart windows and video displays [57, 58].

The ability of Prussian Blue to catalyze the reduction of oxygen (O_2) has been reported too [59]. The reduction of molecular oxygen, also known as ORR (oxygen reduction reaction), has received special attention, given its importance in fuel cell and metal–air batteries. Notwithstanding, the kinetics of the ORR are normally very slow or occur with a very large overpotential, so the presence of a catalyst is essential to speed the process up [60].

Also, in the last three decades, the conception of using Prussian Blue as an electrode material for electrochemical energy storage has arisen. The open channels that the structure exhibits along with the transition metals present on it (Fe^{3+} and Fe^{II}) enable the ionic conductivity and the redox process, as Neff already demonstrated [26, 56].

With the above-exposed examples of possible applications, the multifaceted nature of Prussian blue complexes is evident. However, the subsequent chapters will be focused on the utilization of Prussian Blue and its analogues as active materials for electrochemical energy storage applications.

References

1. L. Samain, F. Grandjean, G.J. Long, P. Martinetto, P. Bordet, D. Strivay, J. Phys. Chem. C **117**, 9693–9712 (2013)
2. A. Ludi, Descriptive chemistry of mixed-valence compounds, in *Mixed-Valence Compounds. NATO Advanced Study Institutes Series (Series C—Mathematical and Phsycial Sciences)*, ed. D.B. Brown, vol. 58 (Springer, Dordrecth 1980), pp. 25–47
3. J.A. Davies, C.M. Hockensmith, V.Y. Kukushkin, Y.N. Kukushkin, *Synthetic Coordination Chemistry: Principles and Practice* (World Scientific, Singapore, 1996)
4. G.E. Stahl, *Experimenta, Observations, Animad-versiones, CCC Numero, Chymicae et Physicae* (Berlin, 1731), pp. 280–283
5. *Miscellanea Berolinensia*, I, 377–378 (1710)
6. J. Woodward, Phil. Trans. 15–17 (1724)
7. J. Kirby, D. Saunders, Nat. Gallery Tech. Bull. **25**, 73–99 (2004)
8. L.J.M. Coleby, Ann. Sci. **4**(2), 206–211 (1939)
9. J.F. Keggin, F.D. Miles, Nature **137**, 577–578 (1936)
10. A. Paolella, C. Faure, V. Timoshevskii, S. Marras, G. Bertoni, A. Guerfi, A. Vijh, M. Armand, K. Zaghib, J. Mater. Chem. A **5**, 1891–18932 (2017)
11. X. Wu, C. Wu, C. Wei, L. Hu, J. Qian, Y. Cao, X. Ai, J. Wang, H. Yang, ACS Appl. Mater. Interfaces **8**, 23706–23712 (2016)
12. M.B. Robin, Inorg. Chem. **1**(2), 337–342 (1962)
13. B.M. Chadwick, A.G. Sharpe, Adv. Inorg. Chem. Radiochem. **8**, 83–176 (1966)
14. A.A. Noyes, J. Am. Chem. Soc. **27**(2), 85–104 (1905)
15. A. Ito, M. Suenaga, K. Ono, J. Chem. Phys. **48**(8), 3597–3599 (1968)
16. H.J. Buser, D. Shwarzenbach, W. Petter, A. Ludi, Inorg. Chem. **16**(11), 2704–2710 (1977)
17. F. Herren, P. Fischer, A. Ludi, W. Halg, Inorg. Chem. 956–959 (1980)
18. P. Day, *Molecules Into Materials: Case Studies in Materials Chemistry-Mixed Valency, Magnetism and Super-Conductivity* (World Scientific, Singapore, 2007), pp. 295–296
19. D. Davidson, J. Chem. Educ. **14**(6), 277 (1937)
20. P.R. Bueno, F.F. Ferreira, D. Giménez-Romero, G.O. Setti, R.C. Faria, C. Gabrielli, H. Perrot, J.J. García-Jareño, F. Vicente, J. Phys. Chem. C **112**, 13264–13271 (2008)
21. J.C. Pramudita, S. Schmid, T. Godfrey, T. Whittle, M. Alam, T. Hanley, H.E.A. Brand, N. Sharma, Phys. Chem. Chem. Phys. **16**, 24178–24187 (2014)
22. H.B. Weiser, W.O. Milligan, J.B. Bates, J. Phys. Chem. **46**(1), 99–111 (1942)
23. E. Fluck, W. Kerler, W. Neuwirth, Angew. Chem. Int. Ed. Engl. **2**, 277–287 (1963)
24. R.J.D. Tilley, *Colour and the Optical Properties of Materials: An Exploration of the Relationship Between Light, the Optical Properties of Materials and Colour*, 2nd edn. (John Wiley & Sons, Chichester (UK), 2010), pp. 342–343
25. A. Kraft, Bull. Hist. Chem. **39**(1), 18–25 (2014)
26. D. Ellis, M. Eckhoff, V.D. Neff, J. Phys. Chem. **85**, 1225–1231 (1981)
27. K. Itaya, H. Akahoshi, S. Toshima, J. Electrochem. Soc. **129**, 1498–1500 (1982)
28. J. Yang, H. Wang, L. Lu, W. Shi, H. Zhang, Cryst. Growth Des. **6**(1), 2438–2440 (2006)
29. M.J. Piernas-Muñoz, E. Castillo-Martínez, O. Bondarchuk, M. Armand, T. Rojo, J. Power Sources **324**, 766–773 (2016)
30. M. Hu, J.S. Jiang, Mat. Res. Bull. **46**, 702–707 (2011)
31. L. Wang, J. Song, R. Qiao, L.A. Wray, M.A. Hossain, Y.-D. Chuang, W. Yang, Y. Lu, D. Evans, J.-J. Lee, S. Vail, X. Zhao, M. Nishijima, S. Kakimoto, J.B. Goodenoug, J. Am. Chem. Soc. **137**, 2548–2554 (2015)
32. Y. You, X.-Q. Yu, Y.-X. Yin, K.W. Nam, Y.-G. Guo, Nano Res. **8**, 117–128 (2015)
33. S.N. Ghosh, J. Inorg. Nucl. Chem. **36**, 2465–2466 (1974)
34. X. Wu, W. Deng, J. Qian, Y. Cao, X. Ai, H. Yang, J. Mater. Chem. A **1**, 10130 (2013)
35. L. Samain, B. Gilbert, F. Grandjean, G.J. Long, D. Strivay, J. Anal. At. Spectrom. **28**, 524 (2013)

36. H.B. Weiser, W.O. Milligan, J.B. Bates, J. Phys. Chem. **46**(1), 99–111 (1942)
37. D.B. Brown, D.F. Shriver, Inorg. Chem. **8**(1), 37–42 (1969)
38. E. Reguera, J.F. Bertran, L. Nuñez, Polyhedron **13**, 1619–1624 (1994)
39. P. Gravereau, E. Garnier, A.M. Hardy, Acta Crystallog. **B35**, 2843–2848 (1979)
40. H. Niwa, W. Kobayashi, T. Shibata, H. Nitani, Y. Moritomo, Sci. Rep. **7**, 13225 (2017)
41. H. Lee, Y.-I. Kim, J.-K. Park, J.W. Choi, Chem. Commun. **48**, 8416–8418 (2012)
42. S.J. Reddy, A. Dostal, F. Scholz, J. Electroanal. Chem. **403**, 209–212 (1996)
43. S. Kajiyama, Y. Mizuno, M. Okubo, R. Kurono, S. Nishimura, A. Yamada, Inorg. Chem. **53**, 3141–3147 (2014)
44. J. Lee, S. Kim, J. Yoon, ACS Omega **2**(4), 1653–1659 (2017)
45. A.A. Karyakin, Electroanalysis **13**(10), 813–819 (2001)
46. L. Nagamoottoo, Surface-supported transition metal hexacyanometallate nanostructured films. Chapter 1 (Ph.D. Thesis), p. 8
47. G.B. Seifer, Russ. J. Inorg. Chem. **7**(5), 1208–1209 (1962)
48. G.B. Seifer, Russ. J. Inorg. Chem. **7**(7), 1746–1748 (1962)
49. S.S. Kaye, J.R. Long, J. Am. Chem. Soc. **127**, 6506–6507 (2005)
50. P. Cartraud, A. Cointot, A. Renaud, J. Chem. Soc., Faraday Trans. **1**(77), 1561–1567 (1981)
51. D.F. Thompson, E.D. Callen, Ann. Pharmacother. **38**(1), 1509–1514 (2004)
52. https://www.accessdata.fda.gov/drugsatfda_docs/label/2008/021626s007lbl.pdf
53. D. Davidson, L.A. Welo, J. Phys. Chem. **32**(8), 1191–196 (1928)
54. M. Verdaguer, A. Bleuzen, C. Train, R. Garde, F. Fabrizi de Biani, C. Desplanches, Philos. Trans. Soc. Lond. Ser. A. **357**, 2959 (1999)
55. M. Taguchi, K. Yamada, K. Suzuki, O. Sato, Y. Einaga, Chem. Mater. **17**(11), 4554–4559 (2005)
56. V.D. Neff, J. Electrochem. Soc. **125**, 886–887 (1978)
57. C.M. Lampert, Sol. Energy Mater. **11**, 1–27 (1984)
58. S. Årman, J. New Mat. Electrochem. Syst. **4**(3), 173–179 (2001)
59. K. Itaya, N. Shoji, I. Uchida, J. Am. Chem. Soc. **106**, 3423–3429 (1984)
60. C. Song, J. Zhang, *PEM Fuel Cell Electrocatalysts and Catalysts Layers, Chapter 2 (Electrocatalytic Oxygen Reduction Reaction)* (Springer, London, 2008), pp. 89–134

Chapter 3
Electrochemical Performance of Prussian Blue and Analogues in Aqueous Rechargeable Batteries

Aqueous batteries offer lower cost compared to non-aqueous batteries and higher ionic mobility, thus potentially higher power, which are the driving forces for their commercialization. On the other hand, water-based electrolytes present a more limited voltage stability window (ca. 1.23 V) than organic electrolytes, what translates into storing lower energy densities than non-aqueous batteries. However, as we will see, there are several ways to enlarge this voltage window, such as modifying the salt chosen and its concentration.

3.1 Initial Considerations: Electrochemistry of Prussian Blue Thin Films

Although Prussian Blue (PB) based thin film batteries have not made it to the market and are less likely to do so than their bulk analogue electrodes, many aspects of the electrochemistry of PB and PBA were initially studied in thin film electrodes and the knowledge was later transferred into bulk PB materials. Therefore, the general aspects of PB thin film batteries will be reviewed here.

The electrochemical behaviour of Prussian Blue thin films was first explored in aqueous electrolyte more than 30 years ago by Neff, as mentioned in Chap. 2 [1]. The thin films were prepared by immersion of a Pt electrode, which had been previously cathodized at constant current in a 1 M HCl solution, into a solution containing 0.01 M $FeCl_2 \cdot 6H_2O$ and 0.01 M $K_3Fe(CN)_6$. After rinsing the electrode and cycling it in a 1 M KCl solution, the reversible cyclic voltammogram (CV) of Fig. 3.1a was achieved. The cathodic voltage swept was accompanied by a change of colour from blue (of the pristine Prussian Blue film) to transparent (of the fully reduced Everitt's salt or Prussian White) and the reaction proposed to take place during such process was the reduction of the Fe^{3+} into Fe^{2+} along with K^+ insertion (see equation Eq. 3.1).

$$KFe^{3+}\left[Fe^{II}(CN)_6\right] + K^+ + e^- \rightarrow K_2Fe^{2+}\left[Fe^{II}(CN)_6\right] \tag{3.1}$$

M. J. Piernas Muñoz and E. Castillo Martínez, *Prussian Blue Based Batteries*, SpringerBriefs in Applied Sciences and Technology, https://doi.org/10.1007/978-3-319-91488-6_3

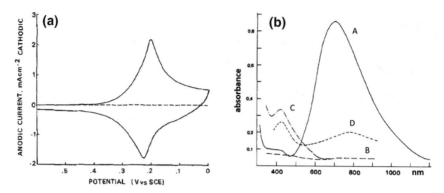

Fig. 3.1 a Single sweep voltammogram of a thin film of Prussian Blue on platinum foil in 1 M KCl solution at 10 mV·sec^{-1} (SCE = saturated calomel electrode). Reproduced with permission from Ref. [1]. Copyright © 1978, Electrochemical Society. **b** Absorption spectra obtained with a SnO$_2$ electrode with 10.5 mC·cm^2 of a PB thin film at different electrode potentials in 1M KCl (pH 4.0): (A) at 0.6 V versus SCE; (B) at −0.2 V; (C) at 1.4 V; (D) at 1.1 V. Reprinted with permission from Ref. [6]. Copyright © 1986, American Chemical Society

Another change of colour from blue to green was observed upon oxidation, which was ascribed to potassium extraction along with Fe^{II} oxidation (Eq. 3.2). Despite this process involved thin film deterioration, presumably as a result of the Cl$^-$ oxidation and oxide formation in the surface of Pt, later work showed that the oxidation process of PB (Eq. 3.2) forming Berlin Green was reversible. Moreover, further oxidation (Eq. 3.3) can result in yellow films, forming the specie known as Prussian Yellow (PY) [2].

$$KFe^{3+}Fe^{II}(CN)_6 \rightarrow 2/3\ K^+ + 2/3\ e^- + K_{1/3}Fe^{3+}\left[Fe^{II}_{1/3}Fe^{III}_{2/3}(CN)_6\right] \quad (3.2)$$

$$K_{1/3}Fe^{3+}\left(Fe^{II}_{1/3}Fe^{III}_{2/3}(CN)_6\right) \rightarrow 1/3\ K^+ + 1/3\ e^- + Fe^{3+}\left(Fe^{III}(CN)_6\right) \quad (3.3)$$

These readily observable colour changes were also characterized by UV–VIS spectroscopy [3], leading to the early application of Prussian Blue thin films as electrodes in electrochromic devices [4]. As it can be observed in Fig. 3.1b, the PB spectrum (A in Fig. 3.1b) is dominated by the absorbance at 700 nm, which follows Beer–Lambert law for different film thicknesses (here not shown). The spectrum of PW (B in Fig. 3.1b) does not show any absorbance, being consistent with its name. As for the oxidized phases BG and PY (D and C, respectively, in Fig. 3.1b), they exhibited an absorbance band at 420 nm corresponding to PY, and an additional band at 770 nm remanent of the 700 nm band of PB was also observed in BG (D in Fig. 3.1b). Investigations on the electrochromic properties and applications of PB-based materials are still ongoing, but we will not give more details here.

A discussion on the exact composition of the PB thin films ('soluble' KFeFe(CN)$_6$ *versus* 'insoluble' Fe$_4$[Fe(CN)$_6$]$_3$) originated because of the multiple parameters associated with the electrodeposition process [5]. The absence of signal from potassium ions in PB electrodes obtained by several analytical spectroscopic techniques, such as atomic absorption, XPS and Auger spectroscopy [6], together with the shift observed in the absorption band of the cycled versus uncycled electrodes led Itaya to propose that cations could not be extracted upon oxidation of PB to BG (Eq. 3.2) or PY (Eq. 3.3) and that, instead, anions (A^-) should be inserted (Eq. 3.4) [7].

$$Fe_4^{3+}[Fe^{II}(CN)_6]_3 + 3\,A^- \rightarrow 3\,e^- + Fe_4^{3+}[Fe^{III}(CN)_6A]_3 \qquad (3.4)$$

Nevertheless, anion insertion is contradictory with the observed dependence of the thermodynamic potential on the concentration of KCl electrolyte, as we explain below [8]. Considering that both the reduction and the oxidation reactions of PB can be treated as compositional solid solutions [8], the Nernst equation describing the mid-peak potential ($E_{1/2}$) for one of them, for example for the reaction of Eq. 3.1, would be that described in Eq. 3.5; where $\alpha(PB)$ and $\alpha(PW)$ are the activities of Prussian Blue and Prussian White, respectively, and $\alpha(K^+)$ is the activity of the potassium ions in the solution phase adjacent to the film. In agreement with the reactions proposed in Eqs. 3.1 and 3.2, the mid-peak potential decreases by diminishing the potassium ion concentration with a Nernstian slope of 59 mV for both reactions, which perfectly agrees with $n = 1$ (being n = number of electrons involved in the reaction) [8]. However, $E_{1/2}$ does not show a similar dependence on the concentration of the anions.

$$E_{1/2} = E^0 + \frac{RT}{F} \ln \frac{\alpha(PB)\alpha(K^+)}{\alpha(PW)} = E^0 - \frac{0.059}{n} \log \frac{\alpha(PW)}{\alpha(PB)\alpha(K^+)} \qquad (3.5)$$

Further evidence of potassium ejection upon oxidation was achieved by chemical analysis, as potassium was found on clean 0.1 M HCl electrolyte solutions after the oxidation of PB thin films [9].

There are plenty of parameters that influence the electrodeposition of PB and PBA thin films, and their electrochemical activity, but probably the three most critical are [10]: (i) the pH of the solution, which was found to be optimal at pH = 1 for the iron hexacyanoferrate (Fe-HCF), since otherwise Fe^{3+} ions are easily hydrolysed and once coordinated by OH^- are not available for bonding in the framework thus generating $[Fe^{II}(CN)_6]^{4-}$ vacancies; (ii) the deposition potential, that should not be lower than 0.2 V versus SCE (standard calomel electrode) for PB, with $Fe^{3+}[Fe^{III}(CN)_6]$ being reduced and deposited at 0.7 V versus SCE (BG $\Rightarrow$ PB process), and Fe^{3+} being reduced at 0.4 V versus SCE (PB $\Rightarrow$ PW process); and (iii) the nature of the non-redox active cations that occupy the zeolitic cubic sites, with K^+, Cs^+, Rb^+ and NH_4^+ promoting the electrochemical activity from PB to PW and others, such as Na^+, behaving as blocking cations. This blocking effect not only depends on the cation but also on the deposition process, the synthetic method and the PBA type. As examples, PB thin films produced by plasma treatment show improved permeability to sodium ions [11], and no blocking effect is observed in bulk PB electrodes (as it will be discussed later). Similarly, electrodeposited thin films of In-, Co- or Ni-HCF were also permeable to Na^+ insertion [12].

Table 3.1 Formal standard potentials of K^+ ion insertion on metal hexacyanometallates $\left(E_{s,f}^o\right)$, standard potentials of the hexacyanoferrates in aqueous solution (E_{aq}), and the difference ΔE between the standard potentials E_{aq} and $E_{s,f}^o$. Extracted with permission from Ref. [14]. Copyright © 2003, John Wiley and Sons

Solid compound [b]	$E_{s,f}^o$ [c] [V vs. SHE]	E_{aq} [V vs. SHE]	$\Delta E = E_{aq} - E_{s,f}^o$ [V]
$Cr^{3+}[Fe(CN)_6]$	1.005	0.355	−0.650
$Fe^{3+}[Fe(CN)_6]$	1.168	0.355	−0.813
$Al^{3+}[Fe(CN)_6]$	1.151	0.355	−0.796
$Ga^{3+}[Fe(CN)_6]$	1.173	0.355	−0.818
$In^{3+}[Fe(CN)_6]$	1.006	0.355	−0.651
$Ni^{2+}[Fe(CN)_6]$	0.857	0.355	−0.502
$Cu^{2+}[Fe(CN)_6]$	0.982	0.355	−0.627
$Co^{2+}[Fe(CN)_6]$	0.966/1.2 [15]	0.355	−0.611/−0.845
$Cd^{2+}[Fe(CN)_6]$	0.858	0.355	−0.504
$Mn^{2+}[Fe(CN)_6]$	0.811	0.355	−0.456
$Pb^{2+}[Fe(CN)_6]$	0.976	0.355	−0.621
$Ag^+[Fe(CN)_6]$	0.880	0.355	−0.525
$Cr^{2+}[Cr(CN)_6]$	−0.562	−1.140	−0.578
$Mn^{2+}[Cr(CN)_6]$	−0.804	−1.140	−0.336
$Fe^{2+}[Cr(CN)_6]$	−0.860	−1.140	−0.280
$Mn^{2+}[Mn^{1+/2+}(CN)_6]$	−0.712	−1.050	−0.338
$Mn^{2+}[Mn^{2+/3+}(CN)_6]$	0.052	−0.240	−0.292
$Fe^{2+}[Mn^{2+/3+}(CN)_6]$	0.075	−0.240	−0.315
$Cr^{3+}[Mn^{2+/3+}(CN)_6]$	0.352	−0.240	−0.592

Interestingly, the thermodynamic redox properties of the $[Fe(CN)_6]^{3-}$ /$[Fe(CN)_6]^{4-}$ pair in solution are very different from those of the solid. It has been found that nickel surface electrodes derivatized with $[Fe(CN)_6]^{3-}$ /$[Fe(CN)_6]^{4-}$ (via irreversible chemisorption) lead to interfaces where the bound material behaves as a solid phase of the type $Ni[Fe(CN)_6]^-$ with minimal solvent interactions, rather than as an anchored species in solution. This is deduced from the much larger voltage dependence on the cation present in the electrolyte and the smaller dependence on the solvent donor number observed for the $[Fe(CN)_6]^{3-}$/$[Fe(CN)_6]^{4-}$ anchored to nickel compared to that found for the solution of $[Fe(CN)_6]^{3-}$/$[Fe(CN)_6]^{4-}$ [13]. In general, hexacyanometallate ions in the solid state are much stronger oxidants than in aqueous solution, as observed by the difference of standard potentials of both reactions (right column in Table 3.1) [14].

This difference in voltage between the solids and the ions in solution can be rationalized considering that, when $[Fe(CN)_6]^{4-}$ is in aqueous solution, the nitrogen

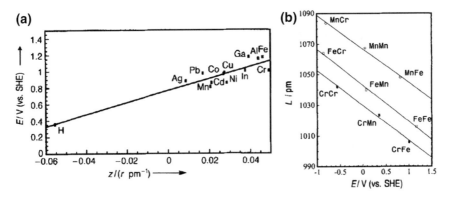

Fig. 3.2 **a** Plot of $E_{s,f}$ of solid metal hexacyanoferrates $M[Fe(CN)_6]$ versus the ratio z/r of the effective charge and radius of the metal ions M including the standard potentials of the hexacyanoferrate ions in water, E_{aq}, as well as the ratio z/r for protons. **b** Plot of the lattice constant (L) of solid hexacyanometallates $M[M'(CN)_6]$ versus their formal potentials of reaction. In this plot, the compounds are indicated as MM'. Reproduced with permission from Ref. [14]. Copyright © 2003, John Wiley and Sons

of the cyanide is coordinated by H^+ from water molecules in the solvent, which have a negative effective ionic radii [16]. Indeed, there is a correlation between the effective radii of the N-coordinated metal (M), its charge and the formal oxidation potential in the solid state (Fig. 3.2a). The more polarizing the cation M (larger charge/ionic radii ratio) typically the higher the voltage of reaction (Fig. 3.2a). This close to linear relationship between voltage and polarizability still holds when including the redox potential of $[Fe(CN)_6]^{4-}$ ions in solution, considering H^+ as M (at the bottom left of the plot). On the other hand, the voltage remains unchanged when the transition metal (TM) replaces part of the interstitial alkali cations [17].

Another conclusion that can be deduced from Table 3.1, and more clearly be observed in Fig. 3.2b, is that the metal M' coordinated by the C of the ligands dominates the redox voltage, whereas the metal M just causes smaller voltage differences. Additionally, for a certain metal M, there is a linear dependence of the lattice parameters with the voltage of reaction on hexacyanochromates (HCCr), hexacyanomanganates (HCMn) and hexacyanoferrates (HCF). This is due to the increase in π-back bonding donation from chromium to iron in the hexacyanometallate ions [14].

3.2 Electrochemistry of Bulk Electrodes: Cathode Materials for Rechargeable Batteries

Although thin film electrodes showed good performance during many cycles, batteries using electrodes thinner than 1 μm would result in not very high capacities. This, along with the fact that the capacity of PB and PBA in the voltage window of water is low per se ($\sim$50–70 mAh·g^{-1} for one redox process) compared to other battery chemistries such as LiCoO$_2$ (140 mAh·g^{-1}) [18], prevented these devices from reaching the market. However, because of their simplicity of synthesis (often at room temperature or close to that), they are of interest for large-scale grid storage applications, provided that bulky electrodes (with higher mass loading) maintain a good performance.

In 2011, when batteries for grid storage became an important focus of attention [19], PB and PBA were revisited. This time to be deployed as bulk electrodes, with the aim of achieving low-cost aqueous batteries with a long cycle life [20]. As the materials' intrinsic electrochemical performance was already known for many PBA, thanks to the studies in thin films, researchers focused on selecting the appropriate material to produce it with good quality and build stable electrodes to enable bulk storage.

As we will see, most of the PBA materials require quite an acidic pH for a stable cycling, which imposes that current collectors are mostly made of carbon, as in Vanadium Redox Flow Batteries (VRFB) [21]. Carbon cloths and graphite felts, which are low-density materials, are especially selected for this purpose.

3.2.1 Materials Choice for Maximizing the Energy Density

There are two ways in which the energy density of the batteries can be increased. One is by increasing the specific capacity of the electrodes. The other is by maximizing the voltage difference between the redox processes at both electrodes, which for cathode materials translates into having the maximum possible voltage of reaction within the electrochemical stability window of the electrolyte.

Figure 3.3a illustrates the Pourbaix diagram of water, with its stability window delimited by the two dashed lines. The yellow area in Fig. 3.3a encompasses the voltages for K$^+$ insertion (or extraction if oxidizing) in the HCF listed in Table 3.1 and shows how most of them have redox voltages close to water oxidation at acidic pHs, while these voltages are above water oxidation at neutral or basic pH. Therefore, acidic pHs are preferred to avoid that H$_2$O oxidation takes place before the redox process associated to HCF. Since only K$^+$ ions are involved in the insertion/extraction reactions and protons are not, the voltage of reduction is pH independent [20]. Details at pH $= 2$ (green rectangle) are zoomed in Fig. 3.3b, manifesting that most of the HCF (except M$' = $ Fe^{3+}, Al^{3+} and Ga^{3+}) could operate at that pH without electrolyte decomposition, which is a good compromise between minimizing the presence of OH$^-$ detrimental to Fe(CN)$_6$ stability and not working at more corrosive acidic pH values. Therefore, the K$^+$ insertion in most of the listed M$'$[Fe(CN)$_6$] solids occurs at adequate voltages to let them operate as cathodes in aqueous batteries. Furthermore, under such conditions of acidic pH, their voltages are high enough to fully utilize the electrochemical stability window of the aqueous electrolyte.

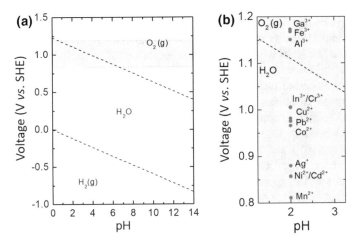

Fig. 3.3 **a** Pourbaix diagram of water. The yellow area indicates the voltage for the reduction reaction of the M´Fe(CN$_6$) materials listed in Table 3.1. The green rectangle is the selected area which is zoomed in **b** and contains the voltages for the most commonly deployed pH $=2$

3.2.1.1 Maximizing Capacity. The Use of Two Redox Processes

Having M-HCF (M = metal) whose $[Fe(CN)_6]^{3-}/[Fe(CN)_6]^{4-}$ redox pair (corresponding to the high voltage process BG $\Rightarrow$ PB) falls within the stability window of the electrolyte is very attractive since it enables the possibility of using 2 electrons per formula unit, as it often happens in organic electrolytes (Chap. 4). The use of both redox processes, BG to PB and PB to ES, would result in a theoretical capacity of ca. 140–200 mAh·g^{-1} (depending on the water content, the metal M and the cation A$^+$ present in the pristine material).

Particularly, the utilization of fully Fe-based compounds is quite appealing since they are the least expensive and the most environmentally friendly. However, it seems that the aqueous electrolyte could only be stable in the presence of BG (Fe[(Fe(CN)$_6$]) at pH ≤ 0.96 (Fig. 3.3a) and such acidic pH is mostly avoided for a practical battery, despite VRFB operate at even lower pH [22]. In fact, attempts to use the full capacity of Fe-HCF electrodes have resulted in capacities as high as 121 mAh·g^{-1} for K$^+$ insertion into crystalline nanosized Prussian Green in KNO$_3$ (pH = 2), with coulombic efficiencies of 98.7% at rates of 1C [23]. Poorer coulombic efficiencies (of 80% at 1C) were achieved during Na$^+$ insertion in 1 M Na$_2$SO$_4$ (pH ~ 7) despite capacities over 100 mAh·g^{-1} were reached [24]. Even with control of the synthesis in ethyleneglycol to produce high-quality dihydrated Prussian White with a low content of vacancies and water, coulombic efficiencies are low and the cathode self-discharges during rest at full charge [25]. Regrettably and notwithstanding having reversible capacities of 120mAh·g^{-1} with a 85% capacity retention over 500 cycles at 21.4C, this low coulombic efficiency would result in largely inefficient batteries.

Other HCFs that in principle could have two usable redox processes inside the electrochemical stability window of aqueous electrolytes at pH = 2 are those based in Co-HCF and Mn-HCF, where the redox process of Co^{3+}/Co^{2+} or Mn^{3+}/Mn^{2+} linked to N is accessible besides that of $[Fe(CN)_6]^{3-}/[Fe(CN)_6]^{4-}$. It has been reported that high-quality rhombohedral nanocrystals of composition $Na_{1.85}Co[Fe(CN)_6]_{0.99}\cdot 2.5H_2O$ have delivered capacities of up to 128 mAh·g^{-1} (~85% of the theoretical 150 mAh·g^{-1}), releasing approximately half of it at each of the two redox processes, which are separated by ~0.6 V. The material retains 84 mAh·g^{-1} at 10C (1C = 130 mA·g^{-1}) and 90% of the capacity after 800 cycles at 5C. However, although coulombic efficiencies above 99% are achieved at high C-rates, at rates of 1C there is still an appreciable coulombic inefficiency likely related to the pH = 7 of the electrolyte [26], which could probably be overcome. When Na-Co-HCF with a higher vacancy content (ca. 14%) is cycled instead, the capacity at 10C drops to approx. 30% of the 100 mAh·g^{-1} achieved at C/2. Nonetheless, this deterioration could also be related to the high concentration of $NaClO_4$ in the electrolyte (10 M) [15]. As for the, Mn-HCF, the Mn^{2+}/Mn^{3+} redox process occurs surprisingly at about 1.4 V versus SHE, i.e., above the voltage of the $[Fe(CN)_6]^{3-}/[Fe(CN)_6]^{4-}$ pair and outside the electrochemical stability of most aqueous electrolytes. The electrolyte oxidation at such a high voltage, together with the phase instability of Mn-HCF and the poor kinetics of Co-HCF, makes the Mn- and Co-containing HCF phases, or their $Co_{1-x}Mn_x$-HCF solid solutions, less likely to be deployed in aqueous full cells, despite the high capacity they can provide [15].

3.2.1.2 Maximizing Voltage: The Use of a Single High Voltage Process

Since Fe and other high capacity materials show coulombic inefficiency issues, the most interesting cathode materials are those that react at slightly lower voltages but are non-toxic and still abundant, such as the Cu^{2+}- and Ni^{2+}-PBA. For these complexes, only the redox pair $[Fe(CN)_6]^{3-}/[Fe(CN)_6]^{4-}$ is supposed to be electrochemically active in aqueous media.

Bulk electrodes of nanostructured copper hexacyanoferrate (Cu-HCF) of composition $K_{0.71}Cu[Fe(CN)_6]_{0.72}\cdot 3.7H_2O$, containing at least 10 mg·cm^{-2}, delivered 60 mAh·g^{-1} (Fig. 3.4a) during K^+ insertion [20]. The material was able to retain 83% of its initial capacity after 40,000 deep discharge cycles at 17C rate with a Qeff of 99.7% (Fig. 3.4b). XRD studies have revealed that during cycling, the material behaves as a solid solution, undergoing a very small isotropic lattice parameter contraction of only 0.9% from the oxidized to the fully reduced phase (Fig. 3.4c) [20]. This very small lattice strain is behind the very long cycle life, as will be discussed later. Also, a key aspect for the performance of the material is that it must be produced by a co-precipitation method to allow the homogeneous crystallization of 20–50 nm large nanoparticles free of impurities. Still, the synthesis takes places at room temperature.

At slightly lower voltages, at 0.59 and 0.69 V versus SHE respectively in 1M $NaNO_3$ and 1M KNO_3 solutions acidified to pH = 2, nanostructured nickel hex-

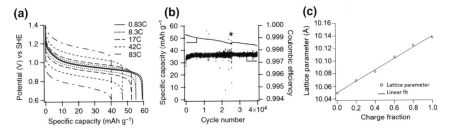

Fig. 3.4 **a** Voltage profile, **b** specific capacity and coulombic efficiency during a life of 40000 cycles and **c** lattice parameter expansion as a function of the amount of K^+ extracted from Cu-HCF in KNO_3 (pH = 2). Reprinted with permission from Ref. [20]. Copyright © 2011, Springer Nature

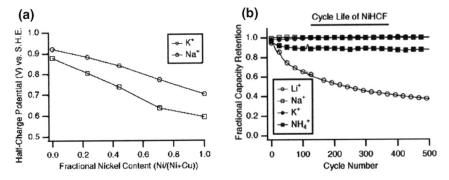

Fig. 3.5 **a** Evolution of the reaction potential of $Cu_{1-x}Ni_x$-HCF with Ni content in ANO_3 (A = Na, K) electrolytes. Reprinted with permission from Ref. [17]. Copyright © 2012, American Chemical Society. **b** Capacity retention of Ni-HCF during cycling of Li^+, Na^+, K^+, and NH_4^+. Reprinted with permission from Ref. [28]. Copyright © 2012, Electrochemical Society

acyanoferrate (Ni-HCF) was able to deliver 52 mAh·g^{-1} of reversible capacity. Notably, no capacity fading was detected after 5,000 cycles at 8.3C (498 mA·g^{-1}) with 1 M $NaNO_3$, due to the extraordinary insertion/de-insertion kinetics in this cation/electrode pair [27].

Optimization of the performance was achieved by producing the $Cu_{1-x}Ni_x$-HCF solid solution. As 'x' augments, the lattice parameter increases and the average voltage for Na^+ or K^+ insertion decreases almost linearly (Fig. 3.5a) [17]. As an example, the average voltage is 144 mV higher for $x = 0.56$ than for $x = 0$. Besides, the phase with x = 0.56 retains 100% of its initial capacity when Na^+ is inserted and 91% in the case of K^+ storage, after 2000 cycles at 500 mA·g^{-1}.

3.2.2 Fast Kinetics and Long Cycle Life in Prussian Blue Analogue

We have already seen that Prussian Blue type materials are not amongst the highest capacity battery electrode materials. Especially in aqueous media often only one redox process is accessible or, if two redox reactions can be accessed, the voltage difference between both of them (~0.6 – 1 V, as it occurs in Co-HCF [26]) is quite large compared to the total voltage of most aqueous cells that would result in low energy density. Indeed, what makes this family of materials outstanding electrodes are two main features: (i) their fast kinetics of reaction, which result in very high coulombic and energy efficiency; and (ii) their very long-term stability upon cycling (up to 10^5 cycles), as it has been just discussed.

3.2.2.1 The Influence of Low Charge Transfer Resistance: Bulk and Thin Film Electrodes

The fast kinetics of reaction of some PBA are a consequence of a low charge transfer resistance. These seem to be related not just to the electrode material nor to the ion being inserted but to the combination of the electrode/ion pair.

In this regard, the influence of several insertion species (K^+, Na^+, Li^+ and NH_4^+) into both nanostructured copper and nickel hexacyanoferrates (Cu-HCF and Ni-HCF) has been directly compared [28]. Rapid kinetics and long cycle life were observed for K^+ insertion into Cu-HCF, and for both Na^+ and K^+ insertion into Ni-HCF and Cu-HCF. Despite Li^+ and NH_4^+ were also intercalated into Ni-HCF, their cyclability and rate capability decayed faster, especially for Li^+ insertion (Fig. 3.5b). This decay was attributed to electrode dissolution, as evidenced by the colour change of the electrolyte solution [28].

The reasons for the dissolution of Cu-HCF and Ni-HCF upon Li^+ insertion, or for the capacity fading upon Na^+ insertion in Cu-HCF, are not known with certainty. Differences in the solubility of the reduced species depending on the electrolyte employed cannot be discarded. However, the large ionic radii of some hydrated species, such as lithium, have been mostly invoked to explain their bad performance in aqueous media [6]. As shown in Table 3.2, the smaller the anhydrous ionic radius, or the larger the charge, the larger the hydrated ionic radii. Considering that the bottleneck radius of cubic PB is 1.6 Å [29], ions with a smaller radius than this value can easily enter without dehydrating. On the other hand, those with a larger ionic radius will probably have to dehydrate to enter the crystal, though they may also rearrange their water coordination sphere from the available hydration water molecules once inside the PB framework.

Specifically, fully hydrated Li^+ ($r_{Li,hydr} = 2.37$ Å) is too large to enter and diffuse through the HCF tunnels. Therefore, Li^+ ions need to be dehydrated to enter the PB structure. Unlike the single redox process typically observed in most monovalent cations, the Li^+ insertion voltage profiles seem to show more than one redox process

Table 3.2 Crystallographic ($r_{M,cr}$) and hydrated ($r_{M,hydr}$) radii (Å) for electrolyte cations and experimental activation energy (E_a) for the charge transfer process in thin Prussian Blue films in organic (PC = propylene carbonate) or aqueous electrolyte. (N.A. = not available)

Cation	$r_{M,cr}$ [16] (Å)	$r_{M,hydr}$ [30] (Å)	E_a (kJ/mol) [31] charge transfer PB/PC	E_a (kJ/mol) [31] charge transfer PB/H$_2$O
Li$^+$	0.9	2.37	52 ± 7	24.1 ± 0.8
Na$^+$	1.02	1.83	51 ± 4	5.1 ± 0.4
K$^+$	1.52	1.38	N.A.	N.A.
Rb$^+$	1.52	1.49	N.A.	N.A.
Cs$^+$	1.67	1.7	N.A.	N.A.
Mg^{2+}	0.86	3.46	N.A.	35.7 ± 0.8

[6, 28], suggesting that Li$^+$ is inserted into multiple sites in PB. In addition to the insertion of lithium into zeolitic sites, it is possible that hydrated lithium ions may be occupying the Fe vacancy sites, as is known to occur with multivalent species (as will be detailed later). In that case, this would indicate that inserted Li$^+$ ions are rehydrated with the interstitial water present in PB, and the movement of these large species through the channels could be directly associated with the material dissolution observed [28]. Although this theory remains to be proven.

The good kinetics of K$^+$ ions in Cu-HCF bulk electrodes can analogously be related with the fact that the size of hydrated K$^+$ is smaller than the bottleneck and therefore it does not need to dehydrate to enter or move inside the channels. Electrochemical Impedance Spectroscopy (EIS) studies showed that the charge transfer resistance of K$^+$ in Cu-HCF is very small (1 ohm/cm^2) [20] and as low as that found on electrodeposited Cu-HCF [31], what allows a fast and reversible ion transport. Since the hysteresis observed between the potential during charge and the potential during discharge increased linearly with the current density, the electrolyte must be the major component to the resistance of the cell, which in a different cell design could be reduced [20].

Besides the low charge transfer resistance, a much lower activation energy for the charge transfer process has been measured for the same inserting species/electrode (thin film) pairs in aqueous solution versus organic electrolyte. The difference in activation energy between organic and aqueous electrolyte is much larger for Na$^+$ ions compared to Li$^+$ ions [29] (see Table 3.2), likely because the size of hydrated Na$^+$ ions ($r_{Na,hydr}$ = 1.83 Å) is closer to the radius of the bottleneck of cubic PB (1.6 Å) than that of Li$^+$ ions. In principle, Na$^+$ ions do not need to completely desolvate in aqueous solution to intercalate into the electrode, in contrast to what occurs when organic electrolyte is used. In fact, ac-electrogravimetry measurements in aqueous media have confirmed that Na$^+$ is partially dehydrated before being inserted into PB [32].

Based on the information described above, it can be concluded that in aqueous electrolyte, cations may need or may not need to be partially or totally dehydrated prior to their insertion into PBA [28–32]. Contrariwise, in organic media, solvated alkali cations would be too large to be inserted into PBA and they all need to be completely desolvated before entering the PB framework structure [29].

The role of aqueous electrolyte in the kinetics of reaction has proven to be important in other electrode materials. This is also true for PB type materials. For instance, in the particular case of Na^+ insertion into PB bulk electrodes, the same PB electrodes containing 10 wt% carbon and 5 wt% PvdF showed much larger overpotential in organic electrolyte (220 mV) at 1C than in aqueous electrolyte (46 mV), which likely results from lower charge transfer resistance [33]. Still, the overpotential in aqueous electrolyte is one order of magnitude smaller for PB than for $NaFePO_4$ [33], suggesting that complete dehydration of Na^+ might be needed in solids 'less porous' than PB and PBA.

3.2.2.2   The Influence of Structural Stability

The other parameters that largely affects the kinetics of reaction and also the cycle life of a material are the degree of volume expansion and the structural distortions. As it will be shown in this section, those PBA electrodes which undergo minimal structural distortions present the lower overpotential, higher energy efficiencies, faster kinetics of reaction and longer cycle life.

Cu-HCF [20], Ni-HCF [27] and their solid solution $Cu_{1-x}Ni_x$-HCF [17] only have one electrochemically active site, Fe^{II}/Fe^{III}-CN, which undergoes minor structural changes during electroactivity. For most of the inserted species, including even divalent and trivalent cations, the materials keep the cubic symmetry, experimenting a minimal lattice contraction (~1%) while cation insertion occurs (Table 3.3). Therefore, this results in very robust mechanical stability during cycling. For instance, Cu-HCF is capable of retaining 83% of the initial capacity after 40000 cycles [20]. Since the tunnels for diffusion of these A^+ cations are large enough to contain most of these ions without distortion, the lattice contractions observed from PBA to PWA (Prussian White analogues) are quite independent of the cation inserted, and these are mostly related to the increase in diameter from 4.295 Å of $[Fe(CN)_6]^{4-}$ to 4.41 Å of $[Fe(CN)_6]^{3-}$ [34]. However, as detailed above, the sphere of coordination of those cations and the solvent at which they are coordinated affect enormously the ease of the A^+ insertion.

In HCF systems with two accessible redox processes, i.e. Fe^{III}/Fe^{II}-CN and M^{3+}/M^{2+}-NC, there is not a smooth single solid solution behaviour in the whole range of cation insertion and phase transitions or phase segregations often take place. PB (or Fe-HCF) is expected to show a transition from cubic to monoclinic or rhombohedral at low voltage upon reduction of the Fe^{III}/Fe^{II}-NC couple, depending on whether K^+ or Na^+ is inserted. Nevertheless, these transformations which have been reported in non-aqueous media [35] (as will be seen in Chap. 4), have not clearly been shown in aqueous media, probably due to the fact that PB with a large number of defects does not necessarily undergo this transition and because the PXRD data in high-quality samples were not conclusive enough [36]. Conversely, the orthorhombic dehydrated

Table 3.3 Lattice contractions upon cation insertion from PBA to PWA and from BGA to PBA in several PBA analogues. All listed materials remain cubic unless the change of symmetry is specified in the same line (being c: cubic, rh: rhombohedral, m: monoclinic, o: orthorhombic)

PBA	A	% Lattice contraction (PBA → PWA)	% Lattice contraction (BGA → PBA)	References
One redox process only				
NiHCF	Na$^+$	–	–	
	K$^+$	1.3	–	[38]
	Mg^{2+}	1.1	–	[38]
	Ba^{2+}	0.9	–	[38]
CuHCF	K$^+$	0.9	–	[20]
	Rb$^+$	0.3	–	[46]
	Pb^{2+}	0.7	–	[46]
Two redox processes				
MnHCF	Na$^+$	~1 (c ⇨ rh)	~−3 (c+c ⇨ c)	[15]
FeHCF	Na$^+$/K$^+$	−(c ⇨ rh/m)		[36]
o-FeHCF	K$^+$	~1.3	~−1.5	[25]
CoHCF	Na$^+$	~2 (c ⇨ rh)	~0	[15]

PW retains orthorhombic symmetry within the whole compositional range [25]. On the other hand, Co-HCF remains cubic in the PB ⇨ BG process but a transition from cubic to rhombohedral occurs upon Na$^+$ insertion and Co^{3+}/Co^{2+}-NC reduction (that is within the PB ⇨ PW process), with a total contraction of about 2–3% [15]. In the case of Mn-HCF, the mechanism is more complex with contributions from both Mn^{3+}/Mn^{2+}-NC and Fe^{III}/Fe^{II}-CN couples in both voltage plateaus, although the Mn^{3+}/Mn^{2+}-NC dominates in the high voltage process and Fe^{III}/Fe^{II}-CN at lower voltages. This complex electronic behaviour is associated with structural changes in Mn-HCF. In the high voltage (HV) process, phase segregation into two cubic phases (Na-rich and Na-poor regions) occurs at the end of oxidation. At low voltages (LV), a cubic to rhombohedral transition takes place accompanied with a noticeable lattice contraction (3%) upon sodium insertion. Since Mn intervenes both in the HV as well as the LV redox processes, both structural features (phase segregation and large lattice contraction) can be related to the change from Jahn–Teller distorted Mn^{3+} to smaller and non-Jahn–Teller Mn^{2+} [15].

In general, we can conclude that the transitions associated with changes in the M^{3+}/M^{2+}-NC redox couple bring along larger changes in lattice parameters and more sluggish kinetics of reaction, as demonstrated by the diffusion coefficients measured by EIS and GITT measurements [15]. As it will be described in the following chapter (Sect. 4.1.1.2), this is not specific of aqueous batteries since it had already been observed in non-aqueous Li-ion batteries.

3.2.3   Multivalent Ion Insertion

Multivalent ions present a larger ionic charge than monovalent ions, which implies bigger hydrated ionic radii (see for example Mg^{2+} vs. Li^+ in Table 3.2) and thus more difficulties to be intercalated into materials without a previous step of dehydration [30]. Insertion of non-solvated ions is also expected to be more hindered that in the case of monovalent cations, given the difficulties to move a larger ionic charge. Moreover, once inside the lattice, the charge balance can require that cations do not accommodate exactly in the same position as monovalent cations or that they keep more water molecules in their hydration shell.

Divalent ions. The insertion of Mg^{2+} into Cu-HCF results in reversible capacities of 50 mAh·g^{-1} at 100 mA·g^{-1} or 60 mAh·g^{-1} at 18 mA·g^{-1} in GITT (galvanostatic intermittent titration technique) mode using a 1 M solution of $Mg(NO_3)_2$ [37]. Such capacity suggests the insertion of 0.3 Mg^{2+} cations considering that magnesium insertion/de-insertion is the only source of capacity, although there is no chemical evidence supporting this. Intriguingly, ^{57}Fe Mossbauer spectroscopy, as well as Cu K-edge XANES data indicate that not only Fe^{III} LS gets reduced to Fe^{II} but also some Cu^{II} gets reduced to Cu^{I} [37]. According to *in situ* XRD, the material behaves as a solid solution [37], as in the case of monovalent cation insertion [20]. However, the insertion expands over a much wider voltage in this case [37] (1 V vs. 0.6 V for K^+ or Na^+) [28], what can be related to its larger hydrated radius and more kinetic hindrance to insertion.

Along with Mg^{2+}, heavier divalent cations with smaller hydrated radii [30], such as Ca^{2+}, Sr^{2+} and Ba^{2+}, have been reversibly inserted into Ni-HCF in 1 M A´$(NO_3)_2$ solutions (A´ = Ca, Sr or Ba) at pH = 2 [38]. Initial capacities of 50 mAh·g^{-1} in a single phase reaction are observed at low C-rates (Fig. 3.6a). At high C-rates, two different redox processes seem to take place (Fig. 3.6a) at the same time that the capacity decreases and the hysteresis increases, especially for Sr^{2+} and Ba^{2+} insertion [38]. The poor capacity retention (inferior to 70% in the first 150 cycles at 5C) observed for Ca^{2+} intercalation can however be largely improved (up to 97%) by the use of superconcentrated electrolyte (8.37 M $Ca(NO_3)_2$) (Fig. 3.6b) [39]. Probably this results from a lower dissolution of the transition metal (TM) or of a less hydrated shell of Ca^{2+} ions or both, as a consequence of the water-poor electrolyte. An alternative strategy to even increase the capacity retention during cycling has been the addition of 20 mM Ni^{2+} in the electrolyte where Ni-HCF is cycled, although the exact mechanism by which the performance improves is not well understood [40].

The intercalation of other divalent cations such as Ni^{2+}, Zn^{2+}, Cu^{2+} or Pb^{2+} in Cu-HCF has also been explored in aqueous media. A comparison of the insertion sites for divalent, monovalent and even trivalent cations will be discussed in the section dedicated to trivalent cations [40]. Since the divalent cation most studied for its insertion in PBA has been Zn^{2+}, given its interest to be used in aqueous zinc batteries, some more details are given below.

Zn^{2+} insertion into PBA. Zn-HCF has naturally been tested as an insertion electrode for Zn^{2+} ions in aqueous media [41]. As described in Chap. 2, Zn-HCF crys-

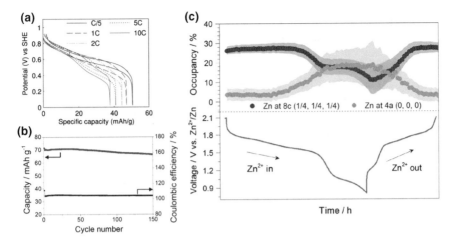

Fig. 3.6 a Voltage profile of Ni-HCF galvanostatically cycled with 1 M Ca(NO$_3$)$_2$ at pH = 2. Reprinted with permission from Ref. [38]. Copyright © 2013, American Chemical Society. **b** Cycle life of Cu-HCF galvanostatically cycled in 8. 37 M Ca(NO$_3$)$_2$. Reprinted with permission from Ref. [39]. Copyright © 2016, The Chemical Society of Japan. **c** Evolution of the Zn occupancy at the zeolitic cavities (8c) and at the [Fe(CN)$_6$]$^{4-}$ vacancy sites (4a) during the first cycle of Zn^{2+} insertion and de-insertion as shown below. Reprinted with permission from Ref. [45] © 2017 The Authors. Published by Elsevier B. V

tallizes in a rhombohedral structure, where Zn cations are located in a tetrahedral environment of N instead of in the octahedral coordination typical of PBA. Actually, particles of 2–3 microns in size with controlled morphology (octahedral, truncated octahedra or cubo-octahedral) could be prepared at room temperature with cubic symmetry, but a slight heating to dry the material converted them into the rhombohedral form while keeping their morphology. When these particles were electrochemically cycled, they showed reversible capacities of 67 mAh·g^{-1} at 1C. Curiously, those particles with cubo-octahedral morphology (less {111} facets than {100} facets in the cubic form) exhibited higher rate performance and longer cycle life than particles of similar size with octahedral morphology (only {111} facets in the cubic form) or truncated octahedral morphology (similar surface of {111} and {100} facets in the cubic form). The fact that the highest rate is achieved for the particles with more {100} planes exposed to the surface agrees with the fact that these {100} planes offer more open tunnels for K$^+$/Zn^{2+} diffusion than the {111} planes [42].

Besides Zn-HCF, the PBA Cu-HCF has also been tested for Zn-ion batteries delivering 90% of the theoretical capacity and a 96% retention after 100 cycles [43]. However, cycling Cu-HCF in a Zn rich electrolyte or at slow C-rates results in an irreversible ageing of the electrode material, which evolves towards a higher voltage at the expense of capacity loss [44]. This ageing has been proposed to involve: first the insertion of Zn^{2+} in vacancy sites and zeolitic sites (which shows up as two different voltage plateaus in the galvanostatic curve, as Fig. 3.6c illustrates), and next a phase segregation into a Cu rich and a Zn rich phases. The sequential Zn^{2+} insertion into

$[Fe(CN)_6]^{4-}$ vacancies was confirmed by *in situ* synchrotron X-ray diffraction on the first cycle [45]. Interestingly, after battery assembly with 1 M $ZnSO_4$ electrolyte, there is a K^+ by Zn^{2+} ion exchange in the zeolitic cavities (8c), decreasing the Zn^{2+} content in this sites as more zinc gets inserted and occupies the vacancies.

Trivalent ions. Even trivalent cations such as Al^{3+} [46], Y^{3+}, La^{3+}, Ce^{3+}, Nd^{3+} or Sm^{3+} have also been reversibly inserted into Cu-HCF, delivering capacities of up to 60 mAh·g^{-1} [40]. Refinement of synchrotron X-ray diffraction patterns of Cu-HCF fully inserted with heavy Rb^+, Pb^{2+} and Y^{3+} has revealed that the preferred site largely depends on the charge of the cation. While more than 90% Rb^+ ions sit in the zeolitic cubic cavity and barely occupy any Fe vacancies, there is a preference for one-third of Pb^{2+} and one-half of Y^{3+} to occupy the $[Fe(CN)_6]^{4-}$ vacancies instead of the cubic A site [40]. This selectivity towards the vacancy site is associated with a larger water content in the unit cell after cation insertion, which suggests that hydration of Pb^{2+} and Y^{3+} helps to shield their large charges and the hydrated cations become stable at these larger vacancy sites [40]. The occupation of these different sites is probably responsible for the appearance of several redox processes in the CV and galvanostatic curves upon insertion of most divalent and trivalent ions, although the contribution of each species to each redox process remains to be assigned yet.

Role of vacancies in multivalent batteries. We have just discussed that multivalent cations prefer to occupy the $[Fe(CN)_6]^{4-}$ vacancy sites rather than the zeolitic sites when they are inserted in PBA. Indeed, it seems that $[Fe(CN)_6]^{4-}$ vacancies are needed for multivalent metal insertion, as vacancy-free high-quality PB and PBA materials, which have shown superior performance in monovalent batteries [26], have not been reported to date as insertion electrodes for divalent and the above-mentioned ions. In addition to this, higher reversible capacities and longer cycle lives have been achieved for multivalent cation storage when small concentrations of Ni^{2+} (20–40 mM) are added into the electrolyte of cells containing Ni-HCF electrodes [38, 40]. Therefore, this suggests that hydrated multivalent ion diffusion through the tunnels may favour TM dissolution, what is however prevented by the presence of the same TM in the electrolyte.

3.3   Full Cells

Full cells based on, at least, one PB or PBA electrode were built as soon as electrodes were developed. Therefore, full batteries based on thin film electrodes as well as those based on bulk electrodes have been reported. In some instances, solid-state batteries were produced without much success, being the flooded cells, i.e. with plenty of electrolyte, the most popular currently. In these full cells, a PB cathode is combined with different types of anode materials, such as other PBA, organic materials (including polymers), inorganic materials or even Zn metal. Amongst the most accepted applications of PBA in aqueous full cells are those in which it is combined with metallic Zn anodes forming a PBA–Zn battery. For these latter, the

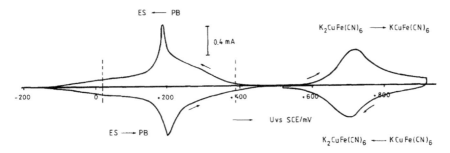

Fig. 3.7 Cyclic voltammograms (CV) of Fe-HCF (left curve; at $v = 20$ mV·s^{-1}) and of Cu-HCF (right curve, $v = 10$ mV·s^{-1}) thin films. Electrolyte: 0.5 M K$_2$SO$_4$; charge density: 9.4 mC·cm^{-1}; electrode area: 0.5 cm^2. Dashed lines show potential limits for cycling of FeHCF. Reprinted with permission from Ref. [48]. Copyright © 1987, Springer Nature

utilization of a Zn^{2+}-based electrolyte, as well as dual batteries with Na$^+$ insertion into a PBA and Zn plating in the anode has mainly been explored.

3.3.1 Full Cells with Thin Films

Already in 1985, Neff built a full cell with two thin films of Prussian Blue deposited in highly porous graphite electrodes. The PB anode would be reduced to Prussian White (or Everitt Salt) during charge, and the PB cathode would oxidize to Berlin Green (with 2/3 of K$^+$ inserted) avoiding the full oxidation to PY [47]. This resulted in a cell of approx. 0.68 V with plenty of challenges ahead, including the adherence of PB to the carbon substrate and capacity fading.

To avoid the oxidation of electrolyte that occurs along with the oxidation process from PB to BG, a cell with two different hexacyanoferrate electrodes that were cycled only in the PB to PW range was assembled. The selected electrodes were K$_2$Cu[Fe(CN)]$_6$ in the cathode and K$_4$Fe$_4$[Fe(CN)$_6$]$_3$ in the anode (Fig. 3.7), whose combination resulted in a total voltage of 0.9 V on full charge and a mean discharge voltage of 0.5 V [48]. Despite not being perfectly balanced, the cell retained 70% of the initial capacity after 766 cycles and a theoretical energy density of 50 Wh·kg^{-1} based on the weight of both electrodes.

3.3.2 Full Cells with Bulky Electrodes

All-solid-state batteries containing a PB cathode and a Cu-HCF or Zn-HCF anode were constructed in the year 2000 [49]. For the electrode preparation, a 15% wt. of the M-HCF powder (M = Fe, Cu or Zn) and 15% wt. KCl were mixed with 70% wt. graphite powder to make a paste that was assembled between a Nafion membrane

Fig. 3.8 Energy densities reported for full cells based on a HCF cathode versus different inorganic and organic anodes. The different chemistries are colour-coded: K-ion (wine), Na-ion (blue), Mg-ion (pink) and Zn-ion (green)

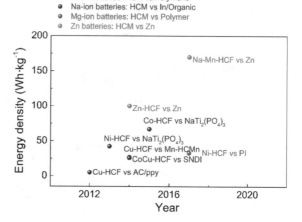

and each of the graphite current collectors. Both full cells (Cu-HCF/PB and Zn-HCF/PB) showed low discharge capacities and voltages which rapidly fell from 0.8 to 0.5 V after 100 cycles. These issues were associated with slow kinetics of electrode reactions, although it is likely that such slow kinetics were due to the high viscosity of the electrolyte and the presence of the Nafion membrane with high ionic resistivity.

In the last decade, flooded cells (also called wet cells) based on different bulk HCF cathodes and organic or inorganic anode materials have also been built. It is important to note that these electrodes have mostly been coated on top of carbon cloth current collectors capable of resisting the acidic pH $= 1 - 2$ of the electrolyte. The evolution of the energy density achieved with the development of these full cells is plotted in Fig. 3.8.

For instance, a full cell based on reduced Cu-HCF as positive electrode and activated carbon/10 wt% reduced polypyrrole anode, with a 1:10 cathode to anode ratio, demonstrated its feasibility. The cell delivered a maximum specific capacity of 54 mAh$\cdot$g$^{-1}_{Cu\text{-}HCF}$ with an average voltage of 0.95 V, that can result in an energy density of 5 Wh$\cdot$kg$^{-1}_{electrode\ materials}$ at 1C rate [50]. Despite the low energy density, the cell showed a 99.9% coulombic efficiency and 95% round trip energy efficiency at 5C with zero capacity loss after 1000 deep discharge cycles.

A big step came from the development of a full cell based on the redox couples NaTi$_2$(PO$_4$)$_3$/Ni-HCF. This cell presented an average output voltage of 1.27 V and a specific energy of 42.5 Wh$\cdot$kg^{-1} but a capacity retention of only 88% after 250 cycles at 5C rate (325 mA$\cdot$g^{-1}) [51]. Another full cell based only on PBA was reported later by Pasta et al., using Cu-HCF as positive electrode and manganese hexacyanoferrate (Mn-HCF) as negative electrode. Such full PB-based cell delivered a maximum specific energy of 27 Wh$\cdot$kg^{-1} at 1C rate with an average voltage of 0.95 V [52]. Furthermore, combining K-CoCu-HCF with sodium naphtalenediimide (SNDI) as organic anode has also led to the construction of a full cell capable of supplying an energy density of 26 $Wh \cdot kg^{-1}$ at a power density of 53 $W \cdot kg^{-1}$ (based on the mass of both active materials) with an average voltage of ~0.9 V [53].

3.3.2.1 Strategies for Opening the Voltage Stability Window in Aqueous Electrolytes

The use of sulphate salts. In a different design, Ni-HCF has also been used as cathode against a polyimide (PI) anode in a full cell with 1 M $MgSO_4$ as electrolyte. The 2-electrode coin cell was operated in a 1.55 V voltage window, which is beyond the thermodynamic stability of water, but no O_2 or H_2 evolution was inferred from the electrochemical data [54]. This less acidic electrolyte has two advantages: (i) carbon felt is not needed as current collector and stainless steel can be deployed; (ii) the electrochemical stability of the electrolyte is typically larger for sulphate-based salts, since the oxidation of water is kinetically hindered. Indeed, Pb–acid batteries (that use H_2SO_4 as electrolyte) also operate in a 2 V window with minor electrolyte decomposition [55], and in carbon-based supercapacitors, Li_2SO_4 is the electrolyte that has enabled the largest stability window in aqueous electrolyte, 1.9 V, with long-term performance [56]. With the $MgSO_4$ electrolyte, the PI/Ni-HCF system was capable to deliver a maximum of 33 $Wh \cdot kg^{-1}$ close to 1 V under a current density of 2 $A \cdot g_{anode}^{-1}$, and showed a capacity retention of 60% after 5000 cycles. Nevertheless, at full charge state, the cell self-discharged a 17% in only 20 h [54]. Even if the system was supposed to operate on Mg^{2+} ion insertion and de-insertion, which is the majority cation in solution, the starting PBA contains Na^+ ions and their activity as insertion cation cannot be ruled out.

Other sulphate salts, such as 1 M $NaSO_4$, have also enabled cycling a $NaTi_2(PO_4)_3$/Co-HCF full cell, with full utilization of the two redox processes of the cathode. This has resulted in one of the largest energy density batteries based on PBA without Zn metal anode. The battery achieved 67 $Wh \cdot kg^{-1}$, and retained 98% of its capacity after 100 cycles [26]. With the same electrolyte, another battery has been built too with a Na-Fe-HCF cathode (using just the PB-ES redox process) and a Zn metal anode in a 1.6 V voltage window [57]. Further details of this cell will be discussed in Sect. 3.3.2.2.

The use of highly concentrated salts. A more extreme case of enlarging the electrochemical stability window comes from the use of highly concentrated salts or water in salt electrolytes, where there are no free water molecules non-coordinated to salt cations or anions and, therefore, they present a different reactivity [58]. For instance, the use of 10 M (17 m) $NaClO_4$ widens the electrochemical stability of the electrolyte to 2.8 V over Ti mesh, enabling the full utilization of the redox processes of a Na-Mn-HCF cathode and a $NaTi_2(PO_4)_3$ anode in a full cell. Despite the apparent stability of the electrolyte, its decomposition is only suppressed at high rates when the capacity retention improves. At current densities of 0.1 $A \cdot g_{cathode}^{-1}$, only 81% of the 117mAh $\cdot g_{cathode}^{-1}$ are retained after 50 cycles, which is much lower than the retentions achieved in $NaNO_3$ or KNO_3 cells at pH $= 1$ [50, 52]. The addition of sodium dodecylsulfate (SDS) and other surfactants employed to avoid metal corrosion have also enabled a 1.7 V Zn/Na-Mn-HCF battery with full utilization of the Mn-HCF capacity. The practicality of this specific cell is based on the ability of the SDS at the critical micelle concentration to suppress Zn anode corrosion and Mn dissolution on the cathode, while widening the electrochemical stability window of the aqueous electrolyte on both regions (water oxidation and reduction) from 1.23 to 2.5 V [59].

3.3.2.2 Prussian Blue–Zn Batteries

Zn insertion into Zn-HCF and others PBA. A Zn/ZnSO$_4$/Zn-HCF full cell has delivered specific capacities of 59.7 mAh·g$^{-1}_{both\ electrodes}$ at an average voltage of 1.7 V, which results in more than 100 Wh·kg^{-1} [41]. However, the cell only retains 76% of its capacity after 100 cycles at C/2 [41]. On the other hand, the substitution of Zn-HCF by Cu-HCF would result in a Zn/ZnSO4/Cu-HCF full cell with an hypothetical energy density of 56Wh·kg [43], which is about half of the value achieved with the Zn-HCF analogue.

Very recently, a biodegradable battery based on a Zn/Zn(OAc)$_2$/Fe-HCF has been proposed to operate with a mixture of choline acetate and water as electrolyte. Conveniently, the presence of ionic liquid in the cell avoids dendrite formation in the Zn side resulting in coulombic efficiencies of 99.5% and energy densities of 50 Wh·kg$^{-1}_{cathode}$ [60]. In light of the improved results, this strategy could open new avenues for the development of Zn/PB cells.

Dual ion insertion batteries. Since cycling M-HCF in Zn rich electrolytes results in the transformation of M-HCF into Zn-HCF, a different approach consists on using a Na-rich electrolyte (1 M NaSO$_4$, 0.01 M H$_2$SO$_4$) [61] or a dual electrolyte (1 M Na$_2$SO$_4$ + 1 M ZnSO$_4$) [59]. These allow the insertion of the preferred monovalent cation in the PB or PBA electrode, while Zn plates and oxidizes on the anode. To completely minimize Zn poisoning at the PB cathode, a battery of high-quality PB cathode (slowly grown to minimize the vacancy content as well) was cycled versus a Zn plate anode with only 1 M Na$_2$SO$_4$ as electrolyte [57]. The full cell delivered 73 mAh·g$^{-1}_{cathode}$ under a current density of 300 mA·g^{-1} at an average voltage of 1.1 V, retaining 80% of the initial capacity after 1000 cycles. Furthermore, if SDS is added to the Na-MnHCF-Zn that employs a mixture of 1 M Na$_2$SO$_4$ and 1 M ZnSO$_4$ as electrolyte [59], the cell is able to deliver 170 Wh·kg^{-1} with a 75% capacity retention after 2000 cycles.

It is evident that HCF-Zn batteries show the largest energy densities amongst all reported PB-containing batteries (Fig. 3.8), due to the low voltage of reaction of Zn metal (-0.76 V vs. SHE) as well as to its relatively high specific capacity ($\sim$819 mAh·g^{-1}). However, and despite their high energy densities, their coulombic efficiency and capacity fading (both associated with electrolyte decomposition at the zinc anode) imply that these batteries are not competitive yet or require further developments before a possible commercialization.

3.4 Viable Commercial Cells Based on Aqueous Prussian Blue and Analogues

The success of PB-based systems is not only reflected in the intense research activities at the academic level but also in the fact that several research groups have embarked on commercial ventures to exploit PBA systems in aqueous rechargeable batteries. As an example, Collin Wessells, from the group at Standford University, was one of the founders of Natron Energy [62] (formerly Alveo Energy), a spin-off company to develop and commercialize sodium-based aqueous batteries based on Prussian Blue analogues. A technology that aims to cover the high rate needs of stationary storage applications.

References

1. V.D. Neff, J. Electrochem. Soc. **125**, 886–887 (1978)
2. K. Itaya, I. Uchida, Inorg. Chem. **25**, 389–392 (1986)
3. K. Itaya, H. Akahoshi, S. Toshima, J. Electrochem. Soc. **129**, 1498–1500 (1982)
4. K. Itaya, H. Akahoshi, S. Toshima, J. App. Phys. **53**, 804–805 (1982)
5. K. Itaya, I. Uchida, V.D. Neff, Acc. Chem. Res. **19**, 162–168 (1986)
6. K. Itaya, T. Ataka, S. Toshima, J. Am. Chem. Soc. **104**, 4767 (1982)
7. K. Itaya, I. Uchida, Inorg. Chem. **25**, 389–392 (1986)
8. D. Ellis, M. Eckhoff, V.D. Neff, J. Phys. Chem. **85**, 1225–1231 (1981)
9. J.W. McCargar, V.D. Neff, J. Phys. Chem. **92**, 3598–3604 (1988)
10. A. Karyakin, Electroanalysis **13**, 10 (2001)
11. A.L. Crumbliss, P.S. Lugg, J.W. Childers, R.A. Palmer, J. Phys. Chem. **89**, 482–488 (1985)
12. S.M. Chen, J. Electroanal. Chem. **521**, 29–52 (2002)
13. A. Bocarsly, S Sinha J. Electroanal. Chem. **140**, 167–172 (1982)
14. F. Scholz, A. Dostal, Angew. Chem. Int. Ed. **34**, 2685–2687 (1995)
15. M. Pasta, R.Y. Wang, R. Ruffo, R. Qiao, H.W. Lee, B. Shyam, M. Guo, Y. Wang, L.A. Wray, W. Yang, M.F. Toney, Y. Cui, J. Mater. Chem. A **4**, 4211–4223 (2016)
16. R.D. Shannon, Acta Cryst **A32**, 751–767 (1976)
17. C.D. Wessells, M.T. McDowell, S.V. Peddada, M. Pasta, R.A. Huggins, Y. Cui, ACS Nano **6**, 1688–1694 (2012)
18. K. Mizushima, P.C. Jones, P.J. Wiseman, J.B. Goodenough, Mat. Res. Bull. **15**, 783–789 (1980)
19. Z. Yang, J. Zhang, M.C.W. Kintner-Meyer, X. Lu, D. Choi, J.P. Lemon, J. Liu, Chem. Rev. **111**, 3577–3613 (2011)
20. C.D. Wessells, R.A. Huggins, Y. Cui, Nat. Commun. **2**, 550 (2011)
21. H. Kaneko, K. Nozaki, Y. Wada, T. Aoki, A. Negishi, M. Kamimoto, Echem. Acta **36**, 1191–1196 (1991)
22. S. Roe, C. Menictas, M. Skyllas-Lazaos, J. Electrochem. Soc. **164**(1), A5023–A5028 (2016)
23. P. Padigi, J. Thiebes, M. Swan, G. Goncher, D. Evans, R. Solanki, Echem. Acta **166**, 32–39 (2015)
24. X. Wu, Y. Luo, M. Sun, J. Quian, Y. Cao, X. Ai, H. Yang, Nano Energy **13**, 117–123 (2015)
25. D. Su, A. McDonagh, S.-Z. Qiao, G. Wang, Adv. Mater. **29**, 1604007 (2017)
26. X. Wu, M. Sun, S. Guo, J. Qian, Y. Liu, Y. Cao, X. Ai, H. Yang, ChemNanoMat. **1**, 188–193 (2015)
27. C.D. Wessells, S.V. Peddada, R.A. Huggins, Y. Cui, Nano Lett. **11**, 5421–5425 (2011)
28. C.D. Wessells, S.V. Peddada, M.T. McDowell, R.A. Huggins, Y. Cui, J. Electrochem. Soc. **159**, A98–A103 (2012)

29. Y. Mizuno, M. Okubo, E. Hoson, T. Hudo, H. Zhou, K. Oh-ishi, J. Phys. Chem. C **117**, 10877–10882 (2013)
30. D.A. Sverjensky, Geochim. Cosmochim. Acta **65**, 3643–3655 (2001)
31. L.M. Siperko, T. Kuwana, J. Echem. Soc. **130**, 396–402 (1983)
32. J.J. García-Jareño, A. Sanmatías, F. Vicente, C. Gabrielli, M. Keddam, H. Perrot, Echem. Acta **45**, 3765–3776 (2000)
33. A.J. Fernandez-Ropero, M.J. Piernas-Muñoz, E. Castillo-Martínez, T. Rojo, M. Casas-Cabanas, Echem. Acta **210**, 352–357 (2016)
34. A. Dostal, G. Kauschka, S.J. Reddy, F. Scholz, J. Electroanal. Chem. **406**, 155–163 (1996)
35. Y. You, X.-L. Wu, Y.-X. Yin, Y. Guo-Guo, Energy Environ. Sci. **7**, 1643–1647 (2014)
36. L. Zhou, Z. Yang, C. Li, B. Chen, Y. Wang, L. Fu, Y. Zhu, X. Liu, Y. Wu, RSC Adv. **6**, 109340 (2016)
37. Y. Mizuno, M. Okubo, E. Hoson, T. Kudo, K. Oh-ishi, A. Okazawa, N. Kojima, R. Kurono, S.-I. Nishimura, A. Yamada, J. Mat. Chem. A **1**, 13055–13059 (2013)
38. R.Y. Wang, C.D. Wessells, R.A. Huggins, Y. Cui, Nano Lett. **13**, 5748–5752 (2013)
39. C.H. Lee, S.-K. Jeong, Chem. Lett. **45**, 1447–1449 (2016)
40. R.Y. Wang, B. Shyam, K.H. Stone, J.N. Weker, M. Pasta, H.-W. Lee, M.F. Toney, Y. Cui, Adv. Energy Mat. **3**, 1401869 (2015)
41. L. Zhang, L. Chen, X. Zhou, Z. Liu, Adv. Energy Mat. **2**, 1400930 (2014)
42. L. Zhang, L. Chen, X. Zhou, Z. Liu, Sci. Rep. **3**, 18263 (2015)
43. R. Trocoli, F. La Mantia. ChemSusChem 8, 481–485
44. G. Kasiri, R. Trocoli, A.B. Hashemi, F. La Mantia, Echem. Acta **222**, 74–83 (2016)
45. V. Renman, D.O. Ojwang, M. Valvo, C.P. Gómez, T. Gustafsson, G. Svensson, J. Power Sources **369**, 146–153 (2017)
46. S. Liu, G.L. Pan, G.R. Li, X.P. Gao, J. Mater. Chem. A **3**, 959–962 (2015)
47. V.D. Neff, J. Electrochem. Soc. **132**, 1382–1384 (1985)
48. E.W. Grabner, S. Kalwellis-Mohn, J. Appl. Electrochem. **17**(3), 653–656 (1987)
49. M. Jayalakshmi, F. Scholz, J. Power Sources **91**, 217–223 (2000)
50. M. Pasta, C.D. Wessells, R.A. Huggins, Y. Cui, Nat. Commun. **3**, 1149 (2012)
51. X. Wu, Y. Cao, X. Ai, J. Quian, H. Yang, Electrochem. Commun. **31**, 145–148 (2013)
52. M. Pasta, C.D. Wessells, N. Liu, J. Nelson, M.T. McDowell, R.A. Huggins, M.F. Toney, Y. Cui, Nat. Commun. **5**, 3007 (2014)
53. D.J. Kim, Y.H. Jung, K.K. Bharathi, S.H. Je, D.K. Kim, A. Coskun, J.W. Choi, Adv. Energy Mater. **4**, 1400133 (2014)
54. L. Chen, J. L. Bao, X. Dong, D. G. Truhlar, Y. -Wang, C. Wang, Y. Xia. ACS Energy Lett. **2**, 1115–1121 (2017)
55. R.S. Treptow, The lead-acid battery: its voltage in theory and practice. J. Chem. Educ. **79**(3), 334–338 (2002)
56. Q. Gao, L. Demarconnay, E. Raymundo-Piñero, F. Beguin, Energy Environ. Sci. **5**, 9611–9617 (2012)
57. L.-P. Wang, P.-F. Wang, T.-S. Wang, Y.-X. Yin, Y.-G. Guo, C.-R. Wang, J. Power Sources **335**, 18–22 (2017)
58. L. Suo, O. Borodin, T. Gao, M. Olguin, J. Ho, X. Fan, C. Luo, C. Wang, K. Xu, Science **350**, 938 (2015)
59. Z. Hou, X. Zhang, X. Li, Y. Zhu, J. Liang, Y. Qian, J. Mat. Chem. A **5**, 730–73 (2017)
60. Z. Liu, P. Bertram, F. Endres, J. Solid State Electrochem. **21**, 2021–2027 (2017)
61. T. Gupta, A. Kim, S. Phadke, S. Biswas, T. Luong, B.J. Hertzberg, M. Chamoun, K. Evans-Lutterodt, D.A. Steingart, J. Power Sources **305**, 22–29 (2016)
62. https://natron.energy/ last accessed on Feb 2018

Chapter 4
Electrochemical Performance of Prussian Blue and Analogues in Non-aqueous High Energy Density Batteries

Although aqueous batteries are more environmentally friendly due to the use of water-based electrolytes, their voltage window is generally limited to the 1.23 V available between the oxidation and reduction of water. Therefore, their utilization in high energy density applications, such as electronic devices and electric vehicles (EV), is hindered. Non-aqueous batteries overcome this limitation by the appropriate election of the electrolyte, where an inorganic salt is dissolved in the selected organic solvents, which typically feature a much wider voltage stability window. As a result, the system can tolerate electrodes with a much larger voltage difference and thus should be capable of storing and releasing higher energy density.

Another advantage, associated to this wider voltage window of the electrolyte, is the possibility of exploiting metallic anodes other than Pb, Cd or Zn used in aqueous batteries. In contrast to aqueous batteries, where the idea of using alkali or alkali-earth metals is inconceivable given their high and dangerous reactivity (exothermic, flammable) with water, non-aqueous batteries could benefit from the low potential and high capacity of these. Nonetheless, such reality is yet to come. So far, and despite the efforts of numerous scientists, dendrites formation along the cycling of the battery and the fatal consequences derived from them (short circuit, heating up and even explosion) cannot be avoided when metallic anodes, or at least alkali metals, are involved. Apparently, this is not the case for alkali-earth metals since no dendrite formation is expected in principle throughout the cycling.

Leaving aside the discrepancies above mentioned and others already discussed in the previous chapter, we will see that many of the phenomena that occur in aqueous batteries also take place in non-aqueous batteries. Some of these common aspects include the dependence of the voltage on the transition metals, the inactivity of some transition metal ions or the important role of the vacancies and water content in the lattice for the electrochemical performance.

© The Author(s) 2018
M. J. Piernas Muñoz and E. Castillo Martínez, *Prussian Blue Based Batteries*,
SpringerBriefs in Applied Sciences and Technology,
https://doi.org/10.1007/978-3-319-91488-6_4

Fig. 4.1 Operation voltages versus first discharge specific capacities of Prussian Blue and analogues systems studied as cathodes for Li-ion batteries. For those materials exhibiting a couple of plateaus, the average voltage plateau has been calculated and is presented in this plot (where $xe^-/f.u. = 'x'$ electrons per formula unit). For additional details and references for each entry, see Table 4.1a

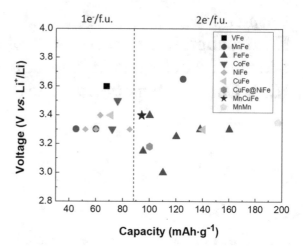

In this chapter, the studies on Prussian Blue materials intended for use in non-aqueous rechargeable batteries are gathered. The information obtained from these is structured first according to the type of cation inserted (Li^+, Na^+, K^+, etc.), i.e. the class of battery (Li-ion, Na-ion, K-ion, etc.) and, within this, based on the Prussian Blue related system.

4.1 Prussian Blue and Analogues in Non-aqueous Li-ion Batteries

4.1.1 Cathode Studies

The majority of the investigations related to the potential application of Prussian Blue and analogues in non-aqueous Li-ion batteries (LIB) have been performed at voltages that are suitable for their use as cathode materials. In the following pages, the most valuable information extracted from the vast majority of them will be reviewed.

4.1.1.1 Bulk Electrodes

According to what was explained in Chap. 1, the higher the operating voltage and the larger the capacity of a material, the higher the energy density it can store and release. The values of these two parameters (voltage and specific capacity), obtained from the electrochemical performance of a wide variety of Prussian Blue systems reported as bulk cathodes for Li-ion batteries, are collected in Fig. 4.1 for comparison.

As it can be observed, the average operating voltage of these materials oscillates between 3.0 and 3.7 V versus Li^+/Li. It is important to note the different voltage scales compared to aqueous batteries (Chap. 3), where the redox voltages were commonly given versus the standard hydrogen electrode (SHE) or standard calomel electrode (SCE) reference electrodes. In this chapter, the counter electrode (which is basically the metal related to the technology under evaluation, e.g. lithium if the technology is LIB) is typically acting also as a reference electrode and, consequently, the voltage is referred to it, unless specified otherwise. Thus, in LIB, the voltage is expressed versus Li^+/Li; whereas for NIB, the voltage will be referred versus Na^+/Na. Remember that these voltages are approx. -3.1 V (Li^+/Li) and -2.9 V (Na^+/Na) versus SHE. Analysing Fig. 4.1, it is clear that hexacyanoferrates containing vanadium and redox active manganese present the higher operating voltage values. As it was advanced in Chap. 3, a redox active metal is a metal that participates actively in the redox reaction and therefore exchanges (accepts o releases) electrons. In future sections, in an analogous way to the previous chapter, we will emphasize which metals are redox active in each case.

As to the gravimetric specific capacity, the Mn–hexacyanomanganate is the system providing the larger capacity, and the iron and copper hexacyanoferrate complexes follow it closely. Nevertheless, within the M–HCF complexes (M = Mn, Fe, Cu), there are remarkable differences in voltage and capacity. Especially for the Mn–HCF and Cu–HCF, in which some materials are able to store one electron per formula unit (f.u.) (as in aqueous electrolyte) and others two. In any case, it is necessary to highlight that the capacity here considered is that obtained in the discharge of the first cycle and that other parameters, such as the coulombic efficiency and the capacity retention, which are discussed through the text, are of great importance to determine the reliability of a material.

Although general trends can be deduced from Fig. 4.1, a more detailed discussion of their electrochemical properties is addressed in upcoming sections.

The Role of Water

In the late 1990s, Imanishi et al. were pioneers in examining the lithium intercalation into *insoluble* Prussian Blue, $Fe_4[Fe(CN)_6]_3 \cdot 14H_2O$, in aprotic media and deciphering the role that different water molecules existing in its lattice play in the lithium insertion mechanism [1]. As it is well known, water can adversely affect the electrochemistry in non-aqueous metal-ion batteries, since it can trigger side reactions [2]. Hence, the importance of finding out its influence on the charge–discharge process. According to Imanishi et al., three types of water coexist in the material: surface-adsorbed water, water physically absorbed in the cubic (or zeolitic) cavities and water structurally coordinated to Fe^{3+}. The removal of surface water and deintercalation of zeolitic water occurs from room temperature to 200 °C, while coordinated-type water can only be eliminated above 200 °C entailing the collapse of the lattice. The comparison of the charge/discharge electrochemical curves of Prussian Blue, previously dried at different temperatures ($\Delta T = 40$–250 °C), revealed that the capacity of the material is strongly conditioned by the number of coordinated water molecules present in it, whereas zeolitic water has little effect. The highest discharge capacity (110 mAh·g^{-1}) being obtained at 3 V versus Li^+/Li for samples dried at 100 and 150 °C, containing a larger amount of coordinated water.

Other Studies on the Fe–Fe System. The Effect of the Alkali Cation, the Stoichiometry of the Species and the Defects Suppression

The electrochemical performance of sodium and potassium hexacyanoferrate nanoparticles (Na-HCF and K-HCF NPs) in LIB was examined by Yu and coworkers [3]. The voltage profiles of both materials displayed a pair of plateaus related to the redox couples Fe^{III}–N/Fe^{II}–N and Fe^{III}–C/Fe^{II}–C, as expected. However, the reaction potential observed for K-HCF was slightly higher than that of Na-HCF (see Table 4.1a). Presumably, such increment results from the partial intercalation of potassium after its extraction in the first charge, which manifests the effect that the alkali cation present in the PB structure has on its electrochemical properties. Nevertheless, the contribution of K^+ decreased upon cycling along with a decline in the average voltage. At a constant current of 5 mA·g^{-1}, the initial discharge capacities obtained were roughly 120 mAh·g^{-1} for Na-HCF and 100 mAh·g^{-1} for K-HCF, with coulombic efficiencies of 97.1% and 96.2%. At higher current densities (100 mA·g^{-1}), more significant differences were appreciated between both materials, especially on their cyclability. K-HCF showed a drastic capacity fading (ca. 35%) over the first 20 cycles in contrast to the excellent capacity retention of Na-HCF that held up 97% of the initial 110 mAh·g^{-1} after 70 cycles. From our point of view, these differences likely arise from the higher crystallinity of Na-HCF particles, as deduced from their PXRD patterns, but the decline of the K^+ contribution to the capacity of K-HCF cannot be ruled out either as another possible cause of failure in the cycling stability.

In addition to the effect of the alkali cation allocated in the structure, the influence of the stoichiometry of iron hexacyanoferrate on the electrochemical properties has also been addressed. In this sense, cubic *'insoluble'* Fe$_4$[Fe(CN)$_6$]$_3$ Prussian Blue and *'soluble'* Berlin Green FeFe(CN)$_6$ nanoparticles (NPs) were evaluated and compared as cathode materials for LIB [4]. Theoretically, Fe$_4$[Fe(CN)$_6$]$_3$ could release a specific capacity of 100 mAh·g^{-1} due to the presence of a single redox active site $\left(Fe_4^{3+}\left[Fe(CN)_6\right] + 4Li^+ + 4e^- \rightarrow Li_4Fe_4^{2+}\left[Fe(CN)_6\right]_3\right)$, while Fe[Fe(CN)$_6$] presents 2 active sites and could transfer 2 electrons $\left(Fe^{3+}\left[Fe^{III}(CN)_6\right] + 2Li^+ + 2e^- \rightarrow Li_2Fe^{2+}\left[Fe^{II}(CN)_6\right]\right)$ reaching capacities of up to 200 mAh·g^{-1}. Under galvanostatic conditions, Fe$_4$[Fe(CN)$_6$]$_3$ and Fe[Fe(CN)$_6$] achieved initial discharge capacities of 95 mAh·g^{-1} (0.95 Li$^+$) and 138 mAh·g^{-1} (1.5 Li$^+$) with coulombic efficiencies higher than 99% after few cycles, which pointed out their high reversibility. In the cyclic voltammograms (CV), two redox peaks were observed upon discharge (at ca. 2.8 and 3.1 V vs. Li$^+$/Li) and along charge (at ca. 3.3 and 3.9 V vs. Li$^+$/Li) for both materials. The presence of two redox processes suggested a slightly different stoichiometry for the insoluble PB, as a single redox active site was expected for this particular species. Anyway, the high voltage plateau corresponds to the redox couple of L.S. Fe^{III}/Fe^{II} coupled to C and the low voltage plateau to the redox pair of H.S. Fe^{3+}/Fe^{2+} bonded to N [5]. In both cases, the integrated area of the upper voltage plateau was by far smaller than that of the lower voltage plateau, a fact that is attributed to the incomplete oxidation of the Fe^{2+} ions linked to C. The two materials displayed reasonable cycling stability, though the rate performance of Fe[Fe(CN)$_6$] was superior to that of the *insoluble* PB, which was

Table 4.1 Representative performance of secondary Li-ion batteries using PBAs (a) bulk electrodes and (b) thin films as cathodes

(a) Bulk electrodes

Active material	Electrode composition (wt.%) (a.m.: C: binder)[a]	Electrolyte[b] (1M salt, solvent mixture)	ΔV (V vs. Li$^+$/Li)	V_{disch} (V vs. Li$^+$/Li)	Q_{disch} (1st cycle) (mAh·g^{-1})	C-rate (mA·g^{-1})	References
$V_4^{3+}[Fe^{2+}(CN)_6]_2$	79.99: 20 AB: 0.1 PTFE	LiClO$_4$, PC: DME	2.5–4.3	3.6	68	0.1 mA·cm^{-2}	[8]
$Mn_3^{2+}[Fe^{3+}(CN)_6]_2$	79.99: 20 AB: 0.1 PTFE	LiClO$_4$, PC: DME	2.5–4.3	3.3	45	0.1 mA·cm^{-2}	[8]
$Na_{4-x}Mn^{2+}[Fe^{II}(CN)_6]_x \cdot zH_2O$	75: 20 AB: 5 PTFE	LiClO$_4$, EC: DEC	2.0–4.3	3.4, 3.9	125	–	[29]
$K_{0.14}Mn_{1.43}^{II}[Fe^{III}(CN)6] \cdot 6H_2O$	75: 20 AB: 5 PTFE	LiClO$_4$, EC: DEC	2.0–4.3	3.3	ca. 60	50	[9]
$Rb_{0.7}Mn_{1.15}^{III}[Fe^{II}(CN)6] \cdot 2.5H_2O$				3.3	ca. 60		
$Fe_4^{3+}[Fe^{II}(CN)_6]_3 \cdot 14H_2O$	79.9: 20 AB: 0.1 PTFE	LiClO$_4$, PC: DME	1.6–4.2	3.0	110	0.1 mA·cm^{-2}	[1]
$Na_xFe^{3+}[Fe^{II}(CN)_6]_y \cdot zH_2O$	70: 15 super P: 15 PVDF	LiPF$_6$, EC: DEC	2.0–4.2	3.0, 3.5	ca. 120	5	[3]
$K_xFe^{3+}[Fe^{II}(CN)_6]_y \cdot zH_2O$				3.0, 3.8	ca. 100		
Soluble $Fe^{3+}[Fe^{III}(CN)_6] \cdot xH_2O$	70: 20 CNTs: 10 PTFE	LiPF$_6$, EC: DMC	1.5–4.0	2.8, 3.8	138	25	[4]
Insoluble $Fe_4^{3+}[Fe^{II}(CN)_6]_3 \cdot yH_2O$				2.7, 3.6	95		
$Fe^{2+}[Fe^{III}(CN)_6]_{0.94}\square_{0.06} \cdot 1.6H_2O$	70 am: 20 C: 10 PTFE	LiPF$_6$, EC: DEC	2.0–4.3	2.9, 3.7	160	0.15 C (24 mA·g^{-1})	[7]
$Co_3^{2+}Fe^{3+}(CN)_6]_2$	79.9: 20 AB: 0.1 PTFE	LiClO$_4$, PC: DME	2.5–4.3	3.3	72	0.1 mA·cm^{-2}	[8]

(continued)

Table 4.1 (continued)

(a) Bulk electrodes

Active material	Electrode composition (wt.%) (a.m.: C: binder)[a]	Electrolyte[b] (1M salt, solvent mixture)	ΔV (V vs. Li$^+$/Li)	V_{disch} (V vs. Li$^+$/Li)	Q_{disch} (1st cycle) (mAh·g^{-1})	C-rate (mA·g^{-1})	References
$Na_{4-x}Co^{2+}[Fe^{II}(CN)_6]_x \cdot zH_2O$	75: 20 AB: 5 PTFE	LiClO$_4$, EC: DEC	2.0–4.3	3.5	76	–	[29]
$Ni_3^{2+}[Fe^{3+}(CN)_6]_2$	79.9: 20 AB: 0.1 PTFE	LiClO$_4$, PC: DME	2.5–4.3	3.3	85	0.1 mA·cm^{-2}	[8]
$Na_{4-x}Ni^{2+}[Fe^{II}(CN)_6]_x \cdot zH_2O$	75: 20 AB: 5 PTFE	LiClO$_4$, EC: DEC	2.0–4.3	3.4	63	–	[29]
$K_xNi^{2+}[Fe^{III}(CN)_6]_y \cdot zH_2O$	75: 20 AB: 5 PTFE	LiClO$_4$, EC: DEC	2.5–4.3	3.3	60	30	[15]
$K_xNi^{2+}[Fe^{III}(CN)_6]_y \cdot zH_2O$	80: 10 AB: 10 PVDF	LiPF$_6$, EC: DEC: DMC	2.0–4.2	3.3	52	0.2 C	[16]
$Cu_3^{2+}[Fe^{3+}(CN)_6]_2$	79.9: 20 AB: 0.1 PTFE	LiClO$_4$, PC: DME	2.5–4.3	3.3	140	0.1 mA·cm^{-2}	[8]
$Na_{4x-2}Cu^{2+}[Fe^{II}(CN)_6]_x \cdot zH_2O$	75: 20 AB: 5 PTFE	LiClO$_4$, EC: DEC	2.0–4.3	3.4	71	–	[29]
$K_{0.1}Cu^{2+}[Fe^{III}(CN)_6]_{0.7} \cdot 3.8H_2O@K_{0.1}Ni^{2+}[Fe^{III}(CN)_6]_{0.7} \cdot 4.1H_2O$	75: 20 AB: 5 PTFE	LiClO$_4$, EC: DEC	2.5–4.3	2.95, 3.24	ca. 100	10	[18]
$K_{0.1}\left[Mn_{0.5}^{2+}Cu_{0.5}^{2+}\right]\left[Fe^{III}(CN)_6\right]_{0.7} \cdot 4H_2O$	75: 20 AB: 5 PTFE	LiClO$_4$, EC: DEC	2.0–4.3	ca. 3.25	94	10	[19]
$K_{1.85}Mn_{1.08}^{2+}\left[Mn^{II}(CN)_6\right] \cdot 0.7H_2O$	–	–	–	–	140	–	[20]
$K_{1.72}Mn^{2+}\left[Mn^{II}(CN)_6\right]_{0.93}\square_{0.07} \cdot 0.65H_2O$	75: 20 AB: 5 PTFE	LiClO$_4$, EC: DEC	2.0–4.3	3.0, 3.7	197	30	[21]

(continued)

Table 4.1 (continued)

(b) Thin films (the electrode composition for all the thin films is 100% a.m.[a] deposited on ITO)

Active material	Electrolyte[b] (1M salt, solvent mixture)	ΔV (V vs. Li$^+$/Li)	V_{disch} (V vs. Li$^+$/Li)	Q_{disch} (1st cycle) (mAh·g^{-1})	C-rate (mA·g^{-1})	References
Li$_x$**Mn**$^{2+}$[Fe^{II}(CN)$_6$]$_{0.81}$·3H$_2$O–ITO	LiClO$_4$, EC: DEC	2.0–4.3	3.5, 3.8	110	1 C	[28]
Li$_x$Mn^{2+}[Fe^{II}(CN)$_6$]$_{0.81}$·3H$_2$O–ITO	LiClO$_4$, EC: DEC	2.0–4.2	3.5, 3.9	110 / 85	2C / 3000C	[27]
Li$_x$Mn^{2+}[Fe^{II}(CN)$_6$]$_{0.83}$·3H$_2$O–ITO	LiClO$_4$, EC: DEC	2.0–4.2	3.5, 3.8	115	0.05 C (6.7 mA·g^{-1})	[21]
Li$_x$Mn^{2+}[Fe^{II}(CN)$_6$]$_{0.87}$·3H$_2$O–ITO	LiClO$_4$, EC: DEC	2.0–4.2		130	0.11 C (14.7 mA·g^{-1})	
Li$_x$Mn^{2+}[Fe^{II}(CN)$_6$]$_{0.93}$·3H$_2$O–ITO	LiClO$_4$, EC: DEC	2.0–4.2		143	0.04 C (6 mA·g^{-1})	
Na$_{1.32}$Mn^{2+}[Fe^{II}(CN)$_6$]$_{0.83}$· 3.5H$_2$O–ITO	LiClO$_4$, EC: DME	2.0–4.2	3.4, 3.6, 3.9	ca. 120	56	[25]
Fe[Fe(CN)$_6$]–ITO (NPB film)	LiClO$_4$, PC	2.04–4.44	2.4, 3.44	10.47 mC·cm^{-2}	200 mV·s^{-1}	[22]
Fe[Fe(CN)$_6$]–ITO (PB film)		2.44–4.44	2.79	5.72 mC·cm^{-2}		
Fe$_4^{3+}$[Fe^{II}(CN)$_6$]$_3$··xH$_2$O	LiPF$_6$, EC: DMC	2.0–4.5	ca. 2.9	980 μC	1 mV·s^{-1}	[23]

(continued)

Table 4.1 (continued)

(b) Thin films (the electrode composition for all the thin films is 100% a.m.[a] deposited on ITO)

Active material	Electrolyte[b] (1M salt, solvent mixture)	ΔV (V vs. Li⁺/Li)	V_{disch} (V vs. Li⁺/Li)	Q_{disch} (1st cycle) (mAh·g⁻¹)	C-rate (mA·g⁻¹)	References
$Li_x Na_{0.13} Co^{2+}[Fe^{II}(CN)_6]_{0.71} \cdot 3.8H_2O$–ITO	LiClO₄, EC: DEC	2.0–4.3	3.5	70	1 C	[28]
$Li_x Na_{0.04} Ni^{2+}[Fe^{II}(CN)_6]_{0.68} \cdot 5.1H_2O$–ITO				65		
$Li_x Na_{0.88} Cd^{2+}[Fe^{II}(CN)_6]_{0.96} \cdot 4.8H_2O$–ITO				68		

[a]a.m. = active material, C = conductive carbon (AB = acetylene black, CNTs = carbon nanotubes); binder (PTFE = polytetrafluoroethylene, PVDF = polyvinylidene fluoride). ITO = indium tin oxide.
[b]PC = propylene carbonate, DME = glyme = 1,2-dimetoxiethane, EC = ethylene carbonate, DEC = diethyl carbonate, DMC = dimethyl carbonate, EMC = ethyl methyl carbonate.

associated to the higher crystallinity and Li^+ diffusion coefficient of $Fe[Fe(CN)_6]$ [4].

Although it has been suggested that the incompatibility between the small radius of Li^+ ($r_{Li}^+ = 0.76$ Å) [6] and the large size of the channels in the PB lattice could be responsible for the commonly observed low capacity and poor cyclability of PB and PBA in LIB, Wu et al. [7] have recently proposed another plausible explanation based on the quantity of structural defects formed during the synthesis. This theory, that has also been discussed for other PBA in aqueous batteries (Chap. 3) and had previously demonstrated in LIB and the NIB counterparts, will be revisited and further argued in future sections of this chapter (Sect. 4.1.1.1 (LIB)—*The Mn–Mn system* and Sect. 4.2.1 (NIB)—*Suppressed vacancies and reduced water content*). In Wu's study, defect-free Berlin Green, $Fe\big[Fe(CN)_6\big]_{0.94}\square_{0.06}\cdot1.6H_2O^1$, with face-centred cubic (fcc) structure (*Fm-3m*), was synthesized. Its cyclic voltammogram (CV) revealed two pairs of redox peaks at 3.7 and 2.9 V versus Li^+/Li during the discharge process, corresponding sequentially to the well-known reduction of LS C–Fe^{III}/Fe^{II} and HS N–Fe^{3+}/Fe^{2+}. In the charge/discharge tests, a reversible capacity of 160 mAh·g^{-1} was delivered at 0.15 C (C = 160 mA·g^{-1}) after the activation of the first cycle, reaching terrific coulombic efficiencies of 99.8%. Furthermore, the formidable amount of 157 mAh·g^{-1} was retained after 100 cycles and, even at 3 C, 90% of the capacity was sustained for 300 cycles. A cyclability that possibly is among one of the best reported among the PBA cathodes for LIB in organic electrolyte. It is evident, therefore, that the suppression of $[Fe(CN)_6]^{4-}$ vacancies supplies a larger amount of redox active centres for Li^+ ion storage and better structural stability, contributing to enhance the electrochemical properties (specific capacity, coulombic efficiency and capacity retention) of PB.

Other PBA systems

As soon as PB bulk electrodes showed activity versus lithium [1], Imanishi and coworkers explored the charge/discharge properties of various Prussian Blue Analogues (PBA) with stoichiometries $M_3^{2+}\big[Fe^{3+}(CN)_6\big]_2$ (M = Mn, Co, Ni, Cu) and $M_4^{3+}\big[Fe^{2+}(CN)_6\big]_3$ (M = V) [8]. All the analogues they synthesized showed face-centred cubic *Fm-3m* symmetry, with slight variations in the lattice parameter on the basis of the metal M substituent. In general, the 'a' parameter decreases as the ionic radius of the M cations increases, except in the case of the vanadium hexacyano-ferrate (V–HCF), whose lattice parameter is similar to that of Prussian Blue. The charge/discharge profiles of all the cyanocomplexes reflected divergences too (see Fig. 4.2). The vanadium complex displayed a discharge potential of 3.6 V versus Li^+/Li ascribed to the reduction of $V^{3+} \rightarrow V^{2+}$, whereas the discharge potentials for the Mn, Co, Ni and Cu analogues were observed around 3.3 V as a result of the reduction of $[Fe^{III}(CN)_6]^{3-} \rightarrow [Fe^{II}(CN)_6]^{4-}$. Regarding the capacity delivered, these materials can be divided into two groups. Those containing V^{3+}, Mn^{2+}, Co^{2+} or Ni^{2+}, which exhibited a capacity between 60 and 80 mAh·g^{-1} as a result

---

[1] As it was already defined in Chap. 2, $\square$ = $[Fe^{II}(CN)_6]^{4-}$ vacancies; or in the case of PBA with formula $AM[M'(CN)_6]\cdot xH_2O$, = $[M'(CN)_6]^{n-}$ vacancies (where n = 3 or 4, depending on the oxidation state of M' (III or II)).

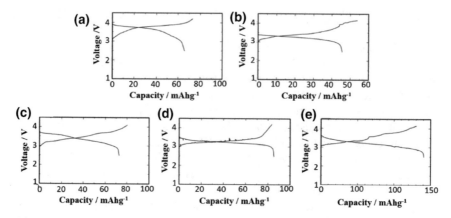

Fig. 4.2 First cycle charge/discharge curves at a current density of 100 mA·cm^{-2} of the hexacyanoferrates: **a** V–Fe, **b** Mn–Fe, **c** Co–Fe, **d** Ni–Fe and **e** Cu–Fe. Reproduced with permission from Ref. [8]. Copyright © 1999 Elsevier Science S.A. All rights reversed

of a single active redox process that involves V for the vanadium complex and Fe for the rest of analogues. And the Cu complex that released almost 140 mAh·g^{-1} and whose high value was attributed to the accessibility of the 2e- redox process, i.e. $[Fe^{III}(CN)_6]^{3-} + e^- \leftrightarrow [Fe^{II}(CN)_6]^{4-}$ and hypothetically $Cu^{2+} + e^- \leftrightarrow Cu^+$. Hence, two main conclusions can be extracted from this investigation: the transition metal replacement lead to substantial changes in the voltage and capacities, and the presence of two active redox pairs enables reaching higher capacities.

The Mn–Fe System

A decade after Imanishi's works, Okubo and collaborators investigated the electrochemical properties of cubic $A_x Mn_y^{2+}[Fe^{III}(CN)_6] \cdot nH_2O(A = K, Rb)$ for Li-ion storage [9]. Almost 1 complete Li$^+$/f.u. could be reversibly inserted into both materials at around 3.3 V versus Li$^+$/Li, providing a capacity of 60 mAh·g^{-1} that was retained during 100 cycles. Nonetheless, even though the voltage, capacity and cyclability of the K- and Rb-containing materials are similar, the Li insertion/extraction mechanism varies from one to another. The combination of XPS and XRD analyses proved that the main redox active site differs between the two compounds, being Fe the active centre for the K salt and Mn the active site for the Rb salt. Apparently, a charge transfer or valence tautomerism between Fe and Mn (from Fe^{III}–CN–Mn^{2+} to Fe^{II}–CN–Mn^{3+}), along with a pressure induced structural change from cubic F-$43m$ to tetragonal I-$4m2$, occurs in the Rb salt during electrode processing.

Additional investigations via soft XAS (X-ray absorption spectroscopy) combined with CTM (charge transfer multiplet) calculations on $K_{0.1}Mn^{2+}[Fe^{III}(CN)_6]_{0.7}$·3.6H$_2$O (Mn–HCF) [10] confirmed that the electronic structure of Mn remains unchanged as HS Mn^{2+} during Li$^+$ insertion/extraction, while Fe atoms are reduced from LS Fe^{III} to LS Fe^{II} upon Li$^+$ insertion (see Fig. 4.3a, b). Besides, a bidirectional charge transfer between the Fe and the CN orbitals, that is MLCT (metal-to-

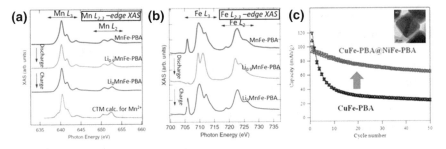

Fig. 4.3 XAS spectra at the **a** Mn $L_{2,3}$ edge and **b** Fe $L_{2,3}$ edge of Li_xMnFe-PBA. A CMT-calculated spectrum for HS Mn^{2+} state is also displayed in (a). Reprinted with permission from D. Asakura, M. Okubo, Y. Mizuno, T. Kudo, H. Zhou, K. Amemiya, F.M.F. de Groot, J.-L. Chen, W.-C. Wang, P.-A. Glans, C. Chang, J. Guo, I. Honma. Physical Review B **84**, 045117 (2011). Copyright © 2011 by the American Physical Society. (c) Cycle stability of the uncoated CuFe–PBA and the CuFe–PBA@NiFe–PBA core@shell particle. Reprinted with permission from Ref. [18]. Copyright © 2013, American Chemical Society

ligand charge transfer) Fe t_{2g} → CN π^* and LMCT (ligand-to-metal charge transfer) CN π/σ → Fe t_{2g}, which suggests a strong electron delocalization, was pointed out as responsible for the stable Li-ion storage properties of Mn–HCF.

The cation A^+ (A = Li, Na) transfer at the electrode–electrolyte interface of $K_{0.1}Mn^{2+}[Fe^{III}(CN)_6]_{0.7} \cdot 3.6H_2O$ was also analysed by EIS (electrochemical impedance spectroscopy) [11]. According to previous research [12, 13], the A^+ transfer can be divided into two processes: the transfer between the Outer Helmholtz plane (OHP) of the electrolyte and the Inner Helmholtz plane (IHP), and the migration between the IHP and the electrode. The former involves the (de)solvation process and the latter the diffusion on the electrode surface. In the case of Li^+ transfer [11], it was found that $E_{a,1}$ (the activation energy invested in the A^+ migration between the OHP and IHP) remains constant in the $0.1 < x < 0.6$ range for Li_xMn–HCF, whereas $E_{a,2}$ (the activation energy for ion diffusion on the electrode surface) depends enormously on 'x'. Note that $0.1 < x < 0.6$ for Li_xMn–HCF implies the (de)insertion of ca. 1 Li^+ per f.u. On the other hand, almost no variation on $E_{a,1}$ (except at $x = 0.6$) and $E_{a,2}$ was noticed when the ion transferred was Na^+. Fact that is consistent with the better rate capability and the invariability of the lattice parameter attained during the Na^+-insertion process [11] compared to the 3.5% volume contraction [9] observed for Li-ion. Again, the effect of the cation inserted in the PB structure is evidenced here.

The Ni-Fe System: A Single-Electron System. Effect of Particle Size on Diffusion
As it was advanced previously, and consistently to what was reported for aqueous batteries, only one e^- redox process takes place in Ni–HCF. Soft XAS combined with CTM calculations on Ni-HCF, $K_{0.1}Ni[Fe(CN)_6]_{0.7}\square_{0.3} \cdot 4.7H_2O$, confirmed the redox activity of Fe and the inactivity of Ni for the storage of Li^+ ions [14]. Analogously to Mn–HCF [10], Ni remained as HS Ni^{2+} during lithium insertion/extraction, while Fe atoms were reduced from LS Fe^{III} to LS Fe^{II} upon Li^+

insertion ($KNi^{II}[Fe^{III}(CN)_6] + xLi^+ + xe^- \rightarrow Li_xKNi^{II}[Fe^{III}_{1-x}Fe^{II}_x(CN)_6]$) [14]. Although, in this case, it was proved that MLCT dominates the electronic configuration of Ni^{2+}–N and Fe^{II}–C bonds, whereas both MLCT and LMCT control the electronic configuration of the Fe^{III}–C bond. Indeed, the higher LMCT contribution on the Fe^{III}–C agrees with having LS Fe^{3+}, which being a d^5 would otherwise prefer to be HS.

The effect of particle size on the storage of Li^+ in NiFe–PBA, $K_xNi[Fe(CN)_6]_y \cdot nH_2O$, was examined as well [15]. With this aim, cubic NiFe–PBA particles of 38, 140, 176, 295 and 387 nm were synthesized and cycled in lithium half-cells. For particles of 176 nm or larger size, a slight boost in the capacity (from 50 or 55 to 60 mAh·g^{-1}) was observed at low cycling rate. And only for the biggest particles (387 nm), a polarization of 0.2 V attributed to a concentration gradient of Li^+ in the NiFe–PBA lattice was appreciated. On the other hand, the rate capability was significantly enhanced with the reduction of the particle size, probably due to the shorter Li^+ diffusion length. In fact, the diffusion coefficient found for Li^+ by GITT (galvanostatic intermittent titration technique) is of the order of 10^{-9}–10^{-8} cm^2·s^{-1} [16], which is greater than most of those reported for other cathodes and agrees with the high rate capability that Ni-HCF can generally stand. However, it must be noted the fact that they did not demonstrate the absence of a contribution from residual K^+ ions, initially contained in the material, to the calculated diffusion coefficient.

The Cu–Fe System. Enhancing the Cycling Stability of a High-Capacity PBA. As it is deduced from Fig. 4.1, some Cu–HCFs are able to insert only one Li^+ ion/f.u. (as it was observed in aqueous systems), while others intercalate two. In this section, we will focus only on those compositions where 2e^--redox processes are available.

According to GITT experiments conducted by Okubo et al., 1.75 Li^+ can be reversibly inserted/extracted from CuFe–PBA, $K_{0.14}Cu^{II}_{1.43}[Fe^{III}(CN)_6] \cdot 5H_2O$ [17]. Since this value exceeds the quantity of Li^+ that can be intercalated by redox reaction of Fe (Fe^{III}/Fe^{II}), they suggested the additional participation of Cu to the reaction, in agreement with Imanishi's assumption [8]. Nonetheless, they did not validate such hypothesis either [17]. Also consistent with Imanishi's result [8], a single plateau was observed in discharge at ca. 3.3 V versus Li^+/Li [17]. Interestingly, the analysis of ex situ XRD patterns of Li_xCuFe–PBA revealed that, at $x < 1.34$, the insertion occurs via solid solution mechanism and the lattice parameter decreases from 10.1337(7) ($x = 0$) to 10.101(6) ($x = 1.34$). However, at $x > 1.34$, the intensity of the initial cubic phase diminishes while a new tetragonal phase emerges, suggesting a biphasic mechanism with phase separation into Li-rich and Li-poor phases, which compromises the cycle life [17].

With the purpose of enhancing the redox stability while simultaneously sustaining high capacity, the same group applied the strategy used in magnetism and prepared a core@shell PBA heterostructure [18]. By combining a high-capacity core $K_{0.1}Cu[Fe(CN)_6]_{0.7} \cdot 3.8H_2O$ (CuFe–PBA) and a lower capacity but a highly stable shell of $K_{0.1}Ni[Fe(CN)_6]_{0.7} \cdot 4.1H_2O$ (NiFe–PBA), they synthesized CuFe–PBA@NiFe–PBA. The comparison of the CVs of CuFe–PBA@NiFe–PBA and bare CuFe–PBA evidenced clear differences between them. The core@shell complex showed sharper cathodic peaks at 3.24 and 2.95 V versus Li^+/Li, corresponding to the sequential reduction of Fe^{3+} and Cu^{2+}, as well as a smaller polarization, that indicates a more reversible and faster redox process than CuFe–PBA. The XAS study of CuFe–PBA@NiFe–PBA confirmed that: (i) there is no change in the Ni^{2+} oxidation state during the whole redox process; (ii) Fe^{3+} is reduced within the $0 < x < 0.7$ range during Li^+ insertion/extraction and does not experience change when $x \geq 0.7$; and (iii) Cu^{2+} remains unchanged in the region $0 < x < 0.7$ and is reduced at $x \geq 0.7$. Curiously, at constant current, both materials displayed only one plateau at 3.3 V versus Li^+/Li. A phenomenon that contrasts with the results observed in the CVs (collected at a slow sweep speed of $0.1\ mV \cdot s^{-1}$) but is in agreement with previous works conducted under constant current conditions [8, 17]. Galvanostatic charge/discharge cycles of CuFe–PBA@NiFe–PBA exhibited lower capacity ($99\ mAh \cdot g^{-1}$) than those of bare CuFe–PBA ($119\ mAh \cdot g^{-1}$), due to the inactivity of Ni^{2+}. Nevertheless, the cycle stability was improved in the core@shell material (see Fig. 4.3c). Stability that was associated with the protective action of the NiFe–PBA shell, which apparently suppresses the cubic to tetragonal phase transition of CuFe–PBA.

Another strategy developed to mitigate the poor cycle stability of CuFe–PBA while maintaining its high capacity consists of the heterometal substitution of the redox active centre Cu^{2+} by redox inactive Mn^{2+}. Through this approach, Okubo and coworkers produced a ternary (MnCu)Fe–PBA [19]. The resulting product, a 40 nm size $K_{0.1}[Mn_{0.5}Cu_{0.5}][Fe(CN)_6]_{0.7} \cdot 4H_2O$, could be indexed as a single cubic phase with a lattice parameter in between of those of MnFe–PBA and CuFe–PBA. During galvanostatic cycling at $30\ mA \cdot g^{-1}$, (MnCu)Fe–PBA delivered an initial discharge specific capacity of $94\ mAh \cdot g^{-1}$ at ca. 3.25 versus Li^+/Li. Such capacity corresponds to the insertion of 0.7 and 0.3 Li^+ respectively associated to the reduction of $[Fe(CN)_6]^{3/4-}$ and $Cu^{2+/+}$ [17, 18], as proved by XPS experiments [19]. Although again the discharge capacity of the ternary analogue was inferior to that of Cu–PBA ($116\ mAh \cdot g^{-1}$), obviously as a result of the redox inactivity of Mn^{2+}, the capacity retention of this ternary material was highly improved.

Thus, it has been demonstrated that partial replacement of redox active metals, such as Cu^{2+}, by inactive redox metals, such as Ni^{2+} or Mn^{2+}, can enhance cycling stability. Although this improvement always occurs at the expense of a decrease in the specific capacity.

The Mn–Mn System

Much less explored than hexacyanoferrates, hexacyanomanganates have also shown interesting electrochemical performance.

When substituting all the Fe by Mn and preparing the Prussian White phase, that is $K_{1.85}Mn_{1.08}[Mn(CN)_6] \cdot 0.7H_2O$ [20], capacities of $140\ mAh \cdot g^{-1}$ are reached. Analogously to the $Cu_3[Fe(CN)_6]_2$ complex reported by Imanishi [8], the high capacity of this material can be explained by means of the two-electron reaction that takes place, involving the Mn^{3+}/Mn^{2+} and $[Mn^{III}(CN)_6]^{3-}/[Mn^{II}(CN)_6]^{4-}$ redox pairs. Nevertheless, as a consequence of the structural phase transition induced by the Jahn–Teller effect of Mn^{3+}, the cyclability of the material is unfortunately poor.

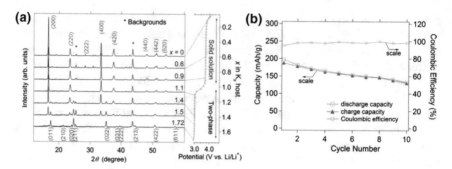

Fig. 4.4 **a** Ex situ XRD patterns of $K_{1.72}Mn[Mn(CN)6]0.93 \cdot \square 0.07 \cdot 0.65H2O$ during K-ion extraction. The red and blue indexes, respectively, correspond to those for the monoclinic and cubic phase. **b** Cyclability and coulombic efficiency of $Mn[Mn(CN)6]0.93\text{-}\square 0.07\text{-}0.65H2O$ versus Li^+/Li during 10 cycles. Reprinted with permission from Ref. [21]. Copyright © 2012, American Chemical Society

As we have discussed before in this chapter and the previous one, the electrochemical behaviour of any material is strongly related to its intrinsic crystal properties. To the best of our knowledge, Asakura and coworkers were the first to relate the $[M'(CN)_6]^{n-}$ vacancies (or $\square$) in PBA (where M' = transition metal, typically Fe; and n = 3 or 4, depending on the oxidation state of M') to the poor coulombic efficiency, the inefficiency for A^+ storage and the possible motif of the structure collapse thereof. To prove it, they cycled again the high-quality monoclinic $(P2_1/n)$ MnMn–PWA $K_{1.85}Mn_{1.08}[Mn(CN)_6] \cdot 0.7H_2O$, as it contained a suppressed amount of vacancies, $K_{1.72}Mn[Mn(CN)_6]_{0.93}\square_{0.07}\cdot 0.65H_2O$ $\left(\square = [Mn(CN)_6]^{4-}\right)$ [21]. This time, K-ions were first electrochemically extracted from the MnMn–PWA framework by GITT, using metallic lithium as counter electrode and a lithium-based electrolyte. Specifically, two processes due to K^+ deintercalation were observed at 3.0 V versus Li^+/Li in the $1.72 > x > 1$ region and at 3.7 V for $x < 1$. At the initial stages of K^+ extraction, $1.72 > x > 1$, the intensities of the monoclinic phase decreased while the reflections of a new cubic phase appeared and increased, indicating that the extraction proceeds via a two-phase process. For $x < 1$, however, a single cubic phase was detected, pointing out that the K^+ extraction at 3.7 V occurs via a solid solution process (see Fig. 4.4a). Consequently, the two redox processes follow a different reaction mechanism. Tracking the CN stretching frequency $\nu(CN)$ by IR, it was deduced that N-coordinated Mn is redox active at 3.0 V and C-bonded Mn is redox active at 3.7 V. The subsequent evaluation of $Mn[Mn(CN)_6]_{0.93}\square_{0.07}\cdot 0.65H_2O$ (i.e. after full extraction of K^+) for the storage of Li^+ delivered an impressive first discharge capacity of 197 mAh·g^{-1} (equivalent to 1.91 Li^+ insertion) at 30 mA·g^{-1} (see Fig. 4.4b). Nonetheless, the material exhibited poor capacity retention. The reaction mechanism and the electronic structure were almost equivalent to those observed during K^+ extraction but with a large volume change (expansion/shrinkage of ca. 17% during the charge/discharge processes), causing a strong mechanical strain in the structure and consequent cracks in the surface of the electrode. The poor cyclability exhibited by Mn–HCMn may, therefore, arise from this volume expansion associated to the cubic $\rightarrow$ monoclinic phase transition, which we assume is owed to the transition from non-distorted Mn^{2+} to Jahn–Teller distorted Mn^{3+}.

Further details about the electrochemical characterization (electrode composition, electrolyte, voltage window, etc.) used for each system described within this 'bulk electrode' section can be found in Table 4.1a. On it, the materials have been organized placing first the HCF complexes and, then, the rest of hexcyanometallates. Within these, PBA have been structured based on the 3d metal(s) that composed them, according to the increase of Z (atomic number). That is, keeping the order of the fourth period of the periodic table (Sc, Ti, V, Cr, Mn, Fe, Co, Ni, Cu, Zn). For example, MnFe systems are described before FeFe systems but after VFe. Finally, in the event of materials containing the same metals, the sequence adopted to present them is: no cation, alkali cations, alkali-earth cations. And, within these, from lower to higher Z, thus pursuing their group order as well. For instance, Li-MnFe is above Na-MnFe but below MnFe.

4.1.1.2 Thin Films

As mentioned in the previous chapter, thin films possess a high surface area which favours the electrochemical reaction. Also, they feature shorter cation diffusion paths compared to bulk electrodes, which translate into good C-rate capabilities. However, PB and PBA thin films do not deliver very high capacities and thus have been less studied for non-aqueous battery applications.

The Fe–Fe System
To assess the durability of Prussian Blue thin films, a 100-nm-thick Prussian Blue film (PB) electrodeposited over bare indium tin oxide (ITO) glass, and a Prussian Blue film electrodeposited on top of a well-dispersed conductive ITO substrate with nanoscale pores (NPB) were prepared and compared [22]. Cracks and poor adhesion between PB and ITO glass were detected in the former when it was observed under SEM (scanning electron microscopy). Conversely, spherical ITO nanoparticles uniformly covered by PB of 40–60 nm size were found in the latter, evidencing not only a higher porosity but also a good adherence and a surface free of cracks. These features were clearly reflected in the CVs, where a rapid decay of the initial charge density ($mA \cdot cm^{-2}$) was detected after 100 cycles for the PB, in contrast to the imperceptible decrease experienced by the NPB over 2000 cycles. Consequently, the better durability of NPB was correlated with the stronger interaction between the PB film and the ITO nanoparticles, highlighting the importance of maintaining the integrity of the material.

Interestingly, the small weight changes that occur during the electrochemical cycling of Prussian Blue thin films could be monitored on a thin film of *insoluble* $Fe_4[Fe(CN)_6]_3$ electrodeposited on a Quartz Crystal Microbalance (QCM) substrate [23]. The CV of PB–QCM showed a cathodic peak at 2.9 V versus Li^+/Li, whose electrical charge exceeded the theoretical capacity elapsed few cycles, prompting that the reduction reaction must include other side reactions in addition to the discharge of PB. According to the QCM results, PB gained weight above 3.1 V versus Li^+/Li during the anodic sweep, which disagrees with the extraction of Li^+ but could be explained by the insertion or adsorption of anionic species for charge compensation, such as for example PF_6^- coming from the electrolyte (see Eq. 4.1). Above 3.5 V, however, the weight of PB decreased, indicating the extraction of Li^+ (see Eq. 4.2). Based on these information, the authors proposed an initial adsorption of PF_6^- during charge which could hinder the posterior extraction of Li^+ and further adsorption of PF_6^-. In the cathodic region, the trend was the opposite to that observed in the anodic process.

$$Li_4Fe_4^{2+}[Fe^{III}(CN)_6]_3 + 3PF_6^- \rightarrow Li_4Fe_4^{2+}[Fe^{III}(CN)_6]_3(PF_6)_3 + 3e^- \quad (4.1)$$

$$Li_4Fe_4^{2+}[Fe^{III}(CN)_6]_3(PF_6)_3 \rightarrow Fe_4^{3+}[Fe^{III}(CN)_6]_3(PF_6)_3 + 4Li^+ + 4e^- \quad (4.2)$$

Undoubtedly, the key finding of this work is that anionic species can be adsorbed in *insoluble* PB, as had already been proposed to occur in aqueous media [24]. Therefore, it is crucial to consider the size of both cationic and anionic species integrating the electrolyte, as well as the pore size of the host structures, to pick out an appropriate combination that does not hinder the redox reaction of the host material.

The Mn–Fe System. Effect of the Electrode Film Thickness and Vacancy Content

Most of the studies focused on evaluating the properties of Mn–HCF thin films as cathode materials for LIB have been conducted by Moritomo's group. During Li^+ intercalation into $Na_{1.32}Mn[Fe(CN)_6]_{0.83} \cdot 3.5H_2O$ thin films (ca. 1 μm) electrode-posited on ITO, three plateaus are distinguished at 3.9, 3.6 and 3.4 V versus Li^+/Li [25]. Among these, the plateau at 3.4 V is ascribed to the reduction of $[Fe^{III}(CN)_6]^{3-}$, and those at higher voltage are attributed to the reduction of Mn^{3+}. Given the redox activity of both transition metals, reversible capacities close to 120 mAh·g^{-1} can be obtained, of which approximately 87% is retained after 100 cycles, probably as a result of the high electronic conductivity of the thin film. This high-capacity value contrasts drastically with that observed in $A_xMn_y^{II}[Fe^{III}(CN)_6]\cdot nH_2O$ bulk electrodes (ca. 60 mAh·g^{-1}) [9]. Nevertheless, such difference can be easily explained considering the different stoichiometry of the bulk electrode $K_{0.1}Mn[Fe(CN)_6]_{0.7} \cdot 3.6H_2O$, where Mn^{2+} is inactive and thus only one redox process ($[Fe(CN)_6]^{3/4-}$) is available.

The effect of the film thickness (L) on the Li^+ insertion into $Li_xMn[Fe(CN)_6]_{0.81} \cdot 3H_2O$ thin film electrodes (Fig. 4.5a, b and c) at different current densities has also been investigated [26]. In this case, only two plateaus at 3.8 and 3.5 V versus Li^+/Li are distinguished, which leads us to think that the third plateau observed previously for the sodiated material ($Na_{1.32}Mn[Fe(CN)_6]_{0.83} \cdot 3.5H_2O$) [25] could be due to the insertion and extraction of Na^+. From the experiments with $Li_xMn[Fe(CN)_6]_{0.81} \cdot 3H_2O$, it was deduced that the thinner the films, the faster current densities they are capable to withstand, since the Li^+ migration length is shorter. For instance, with a

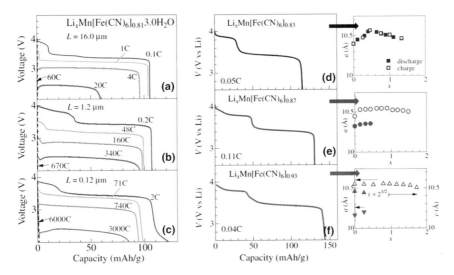

Fig. 4.5 Discharge curves of $Li_xMn[Fe(CN)_6]_{0.81} \cdot 3H_2O$ thin-film electrodes with different thickness: **a** 16 μm, **b** 1.2 μm and **c** 0.12 μm, versus Li^+/Li at various discharge rates. Reproduced from [26]. Copyright © 2012, The Japan Society of Applied Physics. Discharge curve and lattice constant 'a' of: **d** $Li_xMn[Fe(CN)_6]_{0.83} \cdot 3H_2O$ against 'x', **e** $Li_xMn[Fe(CN)_6]_{0.87} \cdot 3H_2O$ against 'x', where filled marks represent the second cubic phase and **f** $Li_xMn[Fe(CN)_6]_{0.93} \cdot 3H_2O$ against 'x', where filled marks represent the second tetragonal phase. Error bar for 'x' is within the symbol size. Reproduced from Ref. [27] with permission. Copyright © 2012, The Japan Society of Applied Physics

$L = 0.12$ μm (Fig. 4.5c), ca. 84.7 mAh·g^{-1} (77% of the initial capacity obtained at 2 C) were retained even at the thundering current density of 3000 C (35 mA·cm^{-2}). Contrariwise, for a $L = 16$ μm (Fig. 4.5a), less than 60% of the specific capacity was held when comparing 20 to 1 C. In general, the fast kinetics of Mn–HCF can be ascribed to the large Li$^+$ diffusion coefficient (D_{Li}^+) and the excellent electrical contact between the active material and the ITO substrate.

In addition to the above exposed, the electrochemical properties of $Li_xMn[Fe(CN)_6]_y \cdot zH_2O$ thin films have been examined as well as a function of the content of vacancies or '$1-y$' [27]. As Fig. 4.5d, e and f illustrate, the discharge capacity increases from 115 ($y = 0.83$) to 130 ($y = 0.87$) and 143 mAh·g^{-1} ($y = 0.93$) as the number of vacancies decreases. Although the capacity delivered by both redox centres (Mn and Fe) was expected to increase with 'y', the capacity associated with the Mn redox process did not follow that trend when comparing '$y = 0.87$' and '$y = 0.93$'. A phenomenon that was explained by structural changes, as reflected in the XRD patterns collected at different states of lithiation 'x'. For $y = 0.87$ and 0.93, a single fcc (*Fm-3m*) phase was appreciated for large Li$^+$ contents, while two phases were differentiated in the region of low lithium content. Interestingly, these two phases are cubic (at $x \leq 0.44$) for $y = 0.87$, whereas a phase separation into a cubic (*Fm-3m*) and a more distorted tetragonal (*I-4m2*) was observed (at $x \leq 0.26$)

for $y = 0.93$. Such phase separation may originate from a spontaneous segregation into cubic Li-rich (Mn^{2+} rich) and tetragonal Jahn–Teller ordered Li-poor (Mn^{3+} rich) domains, with the aim of releasing the local strain caused by the coexistence of Mn^{2+} (with larger ionic radius) and Mn^{3+} (with smaller ionic radius) [27]. In any case, it is thus clear that the structural distortion increases with 'y' when the lithium content 'x' is low.

Other PBA Systems

Other PBA thin film electrodes electrochemically synthesized on ITO for Li^+ intercalation were $(LiNa)_{4y-2}M[Fe(CN)_6]y \cdot zH_2O$ (M = Ni, Co, Mn and Cd) [28]. The voltage profile curves at several C-rates showed discharge plateaus around 3.5 V versus Li^+/Li in all the cases, due to the reduction of Fe (i.e. $[Fe^{III}(CN)_6]^{3-} \rightarrow [Fe^{II}(CN)_6]^{4-}$). An additional plateau was observed only in the Mn analogue at 3.8 V versus Li^+/Li. However, it disappeared at C-rates of 30 C or over, suggesting a slow reduction of Mn^{3+} unable to follow the fast discharge process. Discharge capacities at a moderate rate (1 C) oscillated over 60–70 mAh·g^{-1} for the Ni, Co and Cd complexes, in agreement with the 1-e^- reaction occurring in bulk electrodes [8]. On the other hand, the Mn analogue exhibited ca. 110 mAh·g^{-1} [28], due to the 2-e^- reaction that takes place when M = Mn [25–27]. The materials also exhibited a good C-rate capability, retaining at 100 C up to 74% of the capacity observed at 1 C. To elucidate the origin of the fast Li^+ intercalation that these film electrodes can stand, the authors investigated the D_{Li}^+ of the Mn–HCF by electrochemical impedance spectroscopy (EIS) [28]. The relatively large D_{Li}^+ value, $4 \cdot 10^{-10}$ cm^2/s at $x = 1.1$ and $3 \cdot 10^{-10}$ cm^2·s^{-1} at $x = 0.6$, was attributed to the rigid 3D Li^+ diffusion channel that PBA framework possesses and the good electrical contact between the active material and the ITO substrate. Nonetheless, this D_{Li}^+ does not overcome the 10^{-9}–10^{-8} cm^2·s^{-1} found for Ni–HCF bulk electrodes [16].

In an attempt to contrast the influence that the electrode preparation has on the electrochemical properties, the same group produced $Na_{4x-2}M[Fe(CN)_6]_x \cdot zH_2O$ (M = Mn, Co, Ni and Cu) bulk electrodes and compared them with their corresponding thin films [29], but no substantial differences were found. However, it should not be overlooked the fact that the capacity values obtained for Mn–HCF and Cu–HCF in this case, are notoriously higher or lower than with other bulk electrodes, as Table 4.1 and Fig. 4.1 evidence. The causes prompting this have already been explained for Mn–HCF. As for the Cu–HCF, we believe the reasons are once again merely compositional. Bulk electrodes of stoichiometry $Cu_3^{2+}[Fe^{3+}(CN)_6]_2$ delivered 140 mAh·g^{-1} due to the activity of both $Cu^{2+/+}$ and $Fe^{III/II}$ redox couples [8]. On the other hand, $Na_{4x-2}Cu^{2+}[Fe^{II}(CN)_6]_x \cdot zH_2O$ thin films only supplied 70 mAh·g^{-1}, suggesting that only one of the redox couple must be active (likely Fe^{III}). From our point of view, this could result from preferential (de)intercalation of the Na^+ ions (initially present in the material) versus Li^+ ions, which agrees with the single plateau observed for Cu–HCF when it is evaluated for Na^+ storage, as we will see in future sections.

Influence of the Electrolyte: Organic versus Aqueous

Last but not least, we would like to highlight the significant role that the nature of the electrolyte plays on the electrochemistry. In this sense, Mizuno et al. investigated the electrode–electrolyte charge transfer resistance of Li^+, Na^+ and Mg^{2+} in PBA thin films, aiming to clarify the origin of the higher rate capability supported by these materials when they are cycled in aqueous electrolyte compared to when using organic electrolyte [30]. They concluded that the charge transfer follows different mechanisms depending on the electrolyte used and the cation A^+ ($A^+ = Li^+$, Na^+ or Mg^{2+}) transferred. In general terms, small hydrated cations may not need to be totally or partially dehydrated prior to their insertion into PBA in an aqueous electrolyte, while desolvation is an essential requirement in organic media. For more information, please revisit Chap. 3 (Sect. 3.2.2.1).

Further details about the electrochemical conditions used and the properties achieved for each system described within this 'thin films' section can be found in Table 4.1b.

4.1.2 Anode Studies

As we have just seen, the cathodic properties of Prussian Blue and analogues have been widely investigated, as the more obvious intercalation reactions occur at those voltages. Conversely, there is only a handful of works in which PBA are assessed as anodes for LIB (see Table 4.2). As it will be described within the next few pages, when PBA act as anodes, they present very large reversible capacities due to the utilization of more than one electron per transition metal, although this occurs at the expense of other properties.

The first work that explored Prussian Blue Analogues as negative electrodes in LIBs was published by Shokouhimehr et al. in 2013 [31]. Specifically, nanoparticles of Co-HCF were evaluated in the voltage range 3.0–0.01 V versus Li^+/Li, using 1M $LiPF_6$ in EC: DEC (1: 1% vol) as electrolyte. A first discharge capacity of ca. 900 $mAh \cdot g^{-1}$ with a large irreversible component, that the authors assigned to the electrolyte decomposition and formation of the SEI (Solid Electrolyte Interphase) layer, was observed at a current density of 100 $mA \cdot g^{-1}$. Along the first charge, a very high reversible capacity of 544 $mAh \cdot g^{-1}$ was reached. Ca. 86% of the initial capacity was retained from the 5th cycle to the 30th and, exceptionally, the capacity was still 378 $mAh \cdot g^{-1}$ at a current density of 1000 $mA \cdot g^{-1}$, evidencing the high rate performance of the material also when cycling as an anode.

A year later, Nie and coworkers addressed the utilization of $M_3[Co(CN)_6]_2$ (M = Co, Mn) nanocubes as anodes versus Li^+/Li, using 1M $LiPF_6$ in EC: DMC (1: 1% vol) [32]. In the first discharge of $Co_3[Co(CN)_6]_2 \cdot 11H_2O$, two peaks were distinguished at 1.06 and 0.65 V versus Li^+/Li, which were attributed to the lithiation of the material and the formation of the SEI. In subsequent cycles, however, three redox peaks were observed. The authors speculated that the peak at 1.78/1.88 V was associated to the oxidation/reduction of $N-Co^{3+}/Co^{2+}$, and those at lower voltages (1.23/1.44 and 0.83/1.02 V) would correspond to $C-Co^{III}/Co^{II}$ and $C-Co^{II}/Co^{I}$. During galvanostatic tests at 20 $mA \cdot g^{-1}$, $Co_3[Co(CN)_6]_2 \cdot 11H_2O$ showed a first discharge capacity of 566 $mAh \cdot g^{-1}$, of which 294 $mAh \cdot g^{-1}$ were recovered along the

Table 4.2 Representative performance of secondary Li-ion batteries using PBAs as anodes (Electrode comp.: electrode composition; s.m.: solvent mixture)

Active material	Electrode comp. (wt%) (a.m.: C: binder)[a]	Electrolyte[b] (1M salt, s. m.)	ΔV (V vs. Li$^+$/Li)	V_{ch} (V vs. Li$^+$/Li)	Q_{ch} (1st cycle) (mAh·g^{-1})	C-rate (mA·g^{-1})	References
$Ti_{0.75}Fe_{0.25}[Fe(CN)_6]_{0.96}\cdot 1.9H_2O$	70: 15 AB: 15 PVDF	LiPF$_6$, EC: DEC	0.0–3.0	1.09	350	35	[36]
$Mn^{3+}[Fe^{III}(CN)_6]_{0.6667}\cdot 5H_2O$	70: 20 AB: 10 PVDF	LiPF$_6$, EC: EMC: DMC	0.01–3.0	1.1–1.2	545	200	[33]
$Na_{1.32}Mn[Fe(CN)_6]_{0.84}\cdot 3.4H_2O$	100% a.m. (thin film)	LiClO$_4$, EC: DEC	0.01–3.0	ca. 1.25	370	25 μA·cm^{-2}	[37]
$K_{0.88}Fe^{3+}_{1.04}[Fe^{II}(CN)_6]\cdot 2.83H_2O$	80: 10 C-65: 10 PVDF	LiPF$_6$, EC: DMC	0.005–1.6	ca. 0.9	450/200	9/1750	[34]
$K_x Co^{2+}[Fe^{III}(CN)_6]_y \cdot zH_2O$	70: 15 super P: 15 PVDF	LiPF$_6$, EC: DEC	0.01–3.0	ca. 1.2	544	100	[31]
$Co^{3+}_3[Co^{III}(CN)_6]_2\cdot 11H_2O$	80: 10 AB: 10 PVDF	LiPF$_6$, EC: DMC	0.01–3.0	1.9, 1.4 & 1.0	294.2	20	[32]
$Mn^{3+}_3[Co^{III}(CN)_6]_2\cdot 11H_2O$	80: 10 AB: 10 PVDF	LiPF$_6$, EC: DMC	0.01–3.0	ca. 1.1	257	20	[32]

[a] a.m. = active material, C = conductive carbon (AB = acetylene black, C-65, super P), binder (PVDF = polyvinylidene fluoride).
[b] EC = ethyene carbonate, DEC = diethyl carbonate, EMC = ethyl methyl carbonate, DMC = dimethyl carbonate.

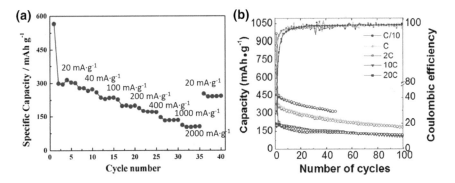

Fig. 4.6 **a** Cycling performance of the $Co_3[Co(CN)_6]_2 \cdot nH_2O$ electrode in the voltage range of 0.01–3.0 V (vs. Li^+/Li) at various current densities. Reproduced from [32] with permission of the Royal Society of Chemistry. **b** Specific capacity of $K_{1-x}Fe_{2+x/3}(CN)_6 \cdot 2.83H_2O$ at different C-rates and their coulombic efficiencies ($1C = 87.5$ mA·g^{-1}/1 Li^+ insertion per f.u.). Reprinted from '$K_{1-x}Fe_{2+x/3}(CN)_6 \cdot yH_2O$, Prussian Blue as a displacement anode for lithium ion batteries', Ref. [34]. Copyright © 2014 with permission from Elsevier

first charge (see Fig. 4.6a). Besides, the material also displayed a satisfactory rate performance, retaining capacities of 132 mAh·g^{-1} at 1 A·g^{-1}. Regarding the other anode, $Mn_3[Co(CN)_6]_2 \cdot nH_2O$, it was barely studied. Despite its huge initial discharge capacity of 868 mAh·g^{-1}, only 355 mAh·g^{-1} of reversible capacity were observed.

In 2015, Xiong synthesized $Mn^{3+}[Fe^{II}(CN)_6]_{0.6667} \cdot 5H_2O$ of 600 nm size [33]. The CV of $Mn^{3+}[Fe^{II}(CN)_6]_{0.6667} \cdot 5H_2O$, collected in the 3.0–0.01 V range versus Li^+/Li, showed one cathodic peak at 1.07 V in the first discharge that disappeared in subsequent cycles, and thus it was attributed to the lithiation of MnFe–PBA and the formation of SEI. From the second cycle onwards, only a symmetric redox peak at 1.24/1.37 V (discharge/charge) versus Li^+/Li was observed. In this case, that peak was attributed to the redox pair Fe^{III}/Fe^{II} coupled to C, even though its voltage is much lower than that at which it has been confirmed to occur (3.3–3.4 V vs. Li^+/Li [8, 9, 29]). The discharge and charge capacities in the first cycle reached 1124 and 545 mAh·g^{-1} at 200 mA·g^{-1}, leading to a coulombic efficiency of 48.5% that increased to 96% after 5 cycles. However, the capacity faded rapidly along the first 30 cycles, and only 300 mAh·g^{-1} (ca. 55%) were retained after 100 cycles.

Conversion Reactions in Prussian Blue and Analogues

Despite the electrochemical performance of PBA as anodes for LIB had already been reported, none of the previous investigations had inquired into the mechanism of reaction that PBA undergo [31, 32].

The first study that undertook such task was reported in 2014 by Piernas et al. [34], who examined the performance of Prussian Blue $K_{1-x}Fe_{2+x/3}(CN)_6 \cdot 2.83H_2O$ ($x = 0.12$) as anode versus Li^+/Li in 1M $LiPF_6$ in EC: DMC (1: 1% vol). In the narrower voltage window of 1.6–0.005 V versus Li^+/Li, the PB nanoparticles exhibited a plateau at ca. 0.9 V during the first discharge and high reversible capacities. The material delivered a charge capacity of 450 mAh·g^{-1} in the first cycle and 350

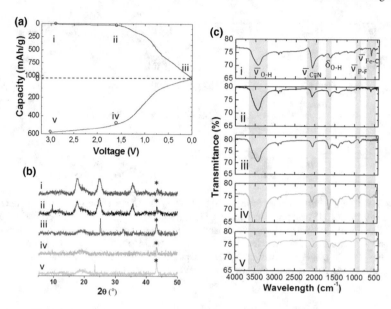

Fig. 4.7 **a** First discharge and charge profile of the Li/PB cell in a voltage range of 0.005–3.2 V. **b** X-ray diffractograms of the electrodes in the different points (*i, ii, iii, iv* and *v*) of the galvanostatic curve of a. The reflections due to the Cu-foil are marked with an*. **c** IR spectra of the electrodes in the different points of the galvanostatic curve. Reprinted from '$K_{1-x}Fe_{2+x/3}(CN)_6 \cdot yH_2O$, Prussian Blue as a displacement anode for lithium ion batteries' [34]. Copyright © 2014 with permission from Elsevier

mAh·g^{-1} after 50 cycles at 8.75 mA·g^{-1} (C/10) (see Fig. 4.6b), which are several times higher than the theoretically expected for the insertion of 1 Li$^+$ per formula unit through an intercalation mechanism. Consequently, some *ex situ* experiments were carried out with electrodes cycled at different stages of charge or discharge to elucidate the reaction mechanism (see Fig. 4.7a). PXRD unveiled the disappearance of the PB reflections once the material was discharged, accompanied by the formation of Fe_2O_3 (Fig. 4.7b). The decrease in the C≡N vibrational band intensity, as well as the cleavage on the Fe-C bond, could be deduced from the *ex situ* IR spectra along the reduction process (Fig. 4.7c). Furthermore, TEM and STEM images detected the presence of Fe-like nanoparticles with interplanar distances of 2.1 Å in the discharged electrode. Based on the previous evidences extracted from the *ex situ* experiments, a conversion (or displacement) mechanism involving metallic Fe^0 nanoparticle formation was proposed, for the first time, as the most plausible reaction taking place when PB was cycled at low voltages versus Li$^+$/Li. In addition to the SEI formation, the reaction detailed in Eq. 4.3 was suggested to occur in the first discharge:

$$K_{1-x}Fe_{1+(x/3)}{}^{3+}Fe^{II}(CN)_6 \cdot yH_2O \xrightarrow[+(1+(x/3))e-]{+(1+(x/3))Li^+} Li_{1+(x/3)}K_{1-x}Fe_{2+(x/3)}{}^{II}(CN)_6 \cdot yH_2O$$

$$\xrightarrow[+2(2+(x/3))e-]{+2(2+(x/3))Li^+} (2+(x/3))Fe^0 + (5+x)LiZ + (1-x)KZ \quad (Z= CN-, OH-, F-,...)$$

$$(4.3)$$

The mechanism occurring upon reoxidation was not determined, though Fe–CN bonds or Fe–O bonds seemed to be reformed, justifying the stable capacity. It is also important to note that even at high current densities, 20 C (1.75 A·g^{-1}), the reversible capacity of the first cycle was as high as 200 mA·g^{-1}, reasserting the good performance of these materials as anodes for LIB. Ongoing *ex situ* XAS and *in situ* ^{57}Fe–Mössbauer experiments are expected to clarify the reaction mechanism of PB anodes both in discharge and charge [35].

Other works exploring the mechanism of Li$^+$ intercalation into PBA in the anodic region were developed by Sun et al. [36] and Moritomo and coworkers [37].

Sun synthesized cubic *Fm-3m* Ti$_{0.75}$Fe$_{0.25}$[Fe(CN)$_6$]$_{0.96}$ · 1.9H$_2$O (Ti–HCF) NPs [36], whose XPS analysis indicated the presence of mainly Fe^{II} and Ti^{IV}, along with a small amount of Fe^{III}. The material was electrochemically tested against metallic lithium and sodium, delivering reversible capacities on the order of 350 and 100 mAh·g^{-1} in the voltage ranges of 0–3.0 V versus Li$^+$/Li and 0–2.5 V versus Na$^+$/Na, respectively. Only the lithium half-cell presented a good cyclability and rate performance, so they focused their efforts on it. According to PXRD performed on cycled electrodes, the cubic structure of PBA disappeared after discharge and did not reappear upon charge, pointing to a conversion reaction mechanism, in agreement with previous results [34]. See Eq. 4.4 for the reaction suggested for the first discharge.

$$Ti_xFe_{2-x}(CN)_6 \cdot 2H_2O + 6Li+ + 6e^- \rightarrow xTi + (2-x)Fe + 6LiCN + 2H_2O$$
$$2H_2O + 4Li + 4e^- \rightarrow H_2 + 2Li_2O \quad (4.4)$$

Further evidence of a conversion mechanism on PBA anodes for LIB was reported by Moritomo's group, who investigated the reaction mechanism of Na$_{1.32}$Mn[Fe(CN)$_6$]$_{0.84}$ · 3.4H$_2$O thin films by XAS and *ex situ* synchrotron X-ray diffraction [37]. XAS data suggested that both metals (Fe and Mn) are reduced to their metallic state upon first discharge, although some residual [Fe(CN)$_6$]$_4$ (ca. 20%) was still present. After charging, however, Mn and Fe are oxidized, being the iron converted into α-Fe$_2$O$_3$. By synchrotron XRD, the formation of Fe metal at the discharged state was confirmed, as well as the formation of α-Fe$_2$O$_3$ and other products (β-Fe$_2$O$_3$, β-FeOOH and γ-FeOOH) at the charged state. No trace of metallic Mn was however detected. A fact that they attributed to its small size. Based on all the information exposed above, they proposed a decomposition reaction of MnFe–PBA followed by the conversion of α-Fe$_2$O$_3$ (see Eq. 4.5), consistent with previous reports [34].

$$\alpha - Fe_2O_3 + 6Li^+ + 6e^- \leftrightarrow 2Fe + 3Li_2O \quad (4.5)$$

Analogously to the cathode section, further details about the electrolyte, voltage window and other electrochemical properties for each PBA system described within this 'anode' section are listed in Table 4.2.

In view of the encouraging results, PBA systems open a new horizon in the exploration of new anode materials for LIB. Although some of the problems to overcome with these materials include the relatively high average voltage plateau (ca. 1 V) and low density, which derives in low gravimetric and volumetric energy density, they benefit from a large capacity and a superior rate capability. PBA can thus be considered as a potential alternative to graphite for LIB in certain applications requiring high power output and fast response. The biggest challenge is probably their first cycle irreversible capacity, which can be slightly improved by the use of PWA instead of PBA [34] but makes precycling almost a must for their utilization. However, if electrode precycling was involved prior to battery assembly, higher capacity anodes such as silicon [38, 39] would probably be preferred.

4.2   Prussian Blue and Analogues as Cathodes in Non-aqueous Na-ion Batteries

Sodium-ion batteries (NIB) are emerging as a strong alternative to LIB. Since the physico-chemical properties of sodium are similar to those of lithium, most of the knowledge and experience acquired for LIB has been directly transferred to NIB, such as the rocking chair concept, many electrodes and electrolyte components and part of the battery assembly process. However, there are substantial differences between the two elements [40]. Namely, the atomic weight of Na is three times heavier than Li, its ionic radio is also larger (1.02 Å vs. the 0.76 Å of lithium) and its standard electrochemical potential is 0.34 V higher. Inevitably, these variations translate into lower gravimetric and volumetric energy density for NIB when contrasted to LIB, and point out the necessity of using host electrode materials with sufficiently large interstitial space to accommodate the bigger Na^+ ions [41]. In this sense, PBA benefit from a 3D framework containing open and tunable channels capable of intercalating the more voluminous Na^+ ions at the same time that can store up to 2 Na^+ per f.u. (and exceptionally 3), which positions them as serious candidates for real market applications.

Unlike in LIB, PBA have been basically explored as cathodes for NIB. Just a couple of works reported the utilization of Ti–HCF [36] and Mn–HCF [37] as anodes for NIB, but they only showed 100 mAh·g^{-1} and no redox activity, respectively. Therefore, in the following pages, we will focus our attention on the performance of the different PBA as cathodes for NIB. Specifically, this section dedicated to NIB has been structured by describing first the electrochemical performance of Prussian Blue phases, then that of Berlin Green derivatives and finally that of the Prussian White phases.

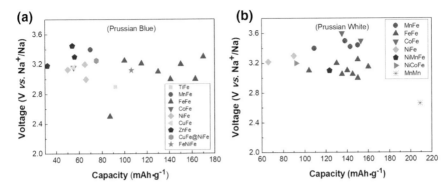

Fig. 4.8 Operation voltages versus first discharge specific capacities of **a** Prussian Blue and **b** Prussian White Analogue systems studied as cathodes for NIB. For those materials exhibiting two plateaus, the average voltage plateau has been calculated and is presented in this plot. In the case of three plateaus, instead of calculating a common average, the percentage of contribution in each one was considered

Figure 4.8 illustrates the average voltage and discharge capacity of a wide variety of Prussian Blue and Prussian White systems tested as cathodes for Na-ion batteries. More experimental details, as well as the corresponding references, are listed in Table 4.3.

Analysing Fig. 4.8, it is easily observed that the operating voltage of these materials typically oscillates between 2.9 and 3.7 V versus Na^+/Na and that the HCF containing Mn, Co and Zn present the higher operating voltage values. As for the gravimetric specific capacity, the Fe–HCF is the clear winner among PBA. For the PWA, however, Mn-hexacyanomanganese (Mn-HCMn) has no rival; while among the HCF, similarly high performance is achieved by Mn–HCF, Co–HCF and Fe–HCF. At this point, we would like to clarify that the low operating voltage observed for the Mn–HCMn results from the wide voltage window in which it was tested (1.3–4.0 V vs. Na^+/Na) and the presence of a third plateau at 1.8 V (as we will see). It is also remarkable the smaller scattering in capacity and voltage that PWA display for each composition, thus manifesting better consistency amongst the results achieved by different authors than in PBA. This must be related to the more stringent synthesis conditions required to produce PWA and a probably closer stoichiometry among different reports. Analogously to LIB, the capacity here considered is that obtained on the first discharge, although other parameters, such as the coulombic efficiency and the capacity retention, are of great importance for determining the reliability of the material.

Table 4.3 Representative performance of secondary Na-ion batteries using PBAs as cathodes. (a) Prussian Blue, (b) Berlin Green and (c) Prussian White

Active material	Electrode composition (wt%) (a.m.: C: binder)[a]	Electrolyte[b] (1M salt, solvent mixture)	ΔV (V vs. Na+/Na)	V_{disch} (V vs. Na+/Na)	Q_{disch} (1st cycle) (mAh·g^{-1})	C-rate (mA·g^{-1})	References
(a) Prussian Blue							
$Na_{0.66}\textbf{Ti}^{3+}[Fe^{II}(CN)_6]_{0.92}\cdot\square_{0.08}$	70: 20 AB: 10 PTFE	NaPF$_6$, EC: DEC	2.0–4.2	2.6, 3.2	92	50	[44]
$Na_{0.7}Ti^{3+}[Fe^{II}(CN)_6]_{0.9}\cdot\square_{0.1}$					82		
$K_x\textbf{Mn}^{2+}[Fe^{III}(CN)_6]$	75: 20 AB: 5 PTFE	NaClO$_4$, EC: DEC	2.0–4.0	3.24, 3.56	ca. 70	C/20	[42]
$K_x\textbf{Fe}^{2+}[Fe^{III}(CN)_6]$	75: 20 AB: 5 PTFE	NaClO$_4$, EC: DEC	2.0–4.0	2.92, 3.58	ca. 100	C/20	[42]
$K_xFe_y^{3+}[Fe^{II}(CN)_6]_z\cdot nH_2O$	70: 25 AB: 5 PTFE	NaClO$_4$, PC	2.0–3.5	ca. 2.5	78	0.125 mA·cm^{-2}	[45]
	70: 25 KB: 5 PTFE			ca. 2.7	87		
$Na_{0.61}Fe^{2+}[Fe^{III}(CN)_6]_{0.94}\cdot\square_{0.06}$	70: 20 KB: 10 PVDF	NaPF$_6$, EC: DEC	2.0–4.2	2.8, 3.3, 3.7, 3.8	170	25	[49]
$Fe_4^{3+}[Fe^{II}(CN)_6]_3$	75: 15 AB: 10 PTFE	NaClO$_4$, PC	1.5–4.0	2.8, 3.2	141	50	[46]
HC–PB/GO	80: 10 AB: 10 PVDF	NaClO$_4$, EC: DMC	2.0–4.2	3.0, 3.4	113 (150 / 2nd cycle)	25	[50]
RGOPC3^c	70: 20 sup P: 10 PVDF	NaClO$_4$, EC: DMC	2.0–4.0	2.7, 3.2–3.4	163.3	30	[53]

(continued)

Table 4.3 (continued)

Active material	Electrode composition (wt%) (a.m.: C: binder)[a]	Electrolyte[b] (1M salt, solvent mixture)	ΔV (V vs. Na$^+$/Na)	V_{disch} (V vs. Na$^+$/Na)	Q_{disch} (1st cycle) (mAh·g^{-1})	C-rate (mA·g^{-1})	References
RGOPC2[c]					152		
RGOPC1[c]					135		
FeHCF NPs@GRs[d]	70: 10 sup P: 10 KB: 10 PVDF	NaClO$_4$, EC: DMC: FEC[g]	2.0–4.2	ca. 2.9, 3.5	150	150	[51]
PB@C[e]	90: 0: 10 PVDF	NaPF$_6$, EC: DEC	2.0–4.0	2.95, 3.4	130 (13th cycle)	50	[52]
OPB1[f]	70: 20 AB: 10 PVDF	NaPF$_6$, EC: DEC	2.0–4.2	2.78, 3.64	110.8	50	[47]
OPB2[f]					115.2		
Na$_{0.75}$Fe$_{2.08}$(CN)$_6$·3.4H$_2$O	80: 10 C-65: 10 PVDF	NaPF$_6$, EC: PC: FEC[h]	2.4–4.2	2.8, 3.4	130	9.2	[58]
K$_x$Co^{2+}[Fe^{III}(CN)$_6$]	75: 20 AB: 5 PTFE	NaClO$_4$, EC: DEC	2.0–4.0	3.16	ca. 55	C/20	[42]
K$_x$Ni^{2+}[Fe^{III}(CN)$_6$]	75: 20 AB: 5 PTFE	NaClO$_4$, EC: DEC	2.0–4.0	3.13	ca. 50	C/20	[42]
K$_{0.09}$Ni^{2+}[Fe^{III}(CN)$_6$]$_{0.71}$·6H$_2$O	80: 10 CB: 10 PVDF	NaPF$_6$, EC: PC	2.0–4.1	ca. 3.0	66	20	[56]
K$_x$Ni^{2+}[Fe^{II}(CN)$_6$]	70: 20 CB: 10 PTFE	NaClO$_4$, EC: PC	2.5–3.8	3.2	65	10	[48]

(continued)

Table 4.3 (continued)

Active material	Electrode composition (wt%) (a.m.: C: binder)[a]	Electrolyte[b] (1M salt, solvent mixture)	ΔV (V vs. Na$^+$/Na)	V_{disch} (V vs. Na$^+$/Na)	Q_{disch} (1st cycle) (mAh·g^{-1})	C-rate (mA·g^{-1})	References
$K_x Cu^{2+}[Fe^{III}(CN)_6]$	75: 20 AB: 5 PTFE	NaClO$_4$, EC: DEC	2.0–4.0	3.16	ca. 55	C/20	[42]
$K_x Zn^{2+}[Fe^{III}(CN)_6]$	75: 20 AB: 5 PTFE	NaClO$_4$, EC: DEC	2.0–4.0	3.18	ca. 32	C/20	[42]
$Na_2 Zn_3^{2+}[Fe^{III}(CN)_6]_2 \cdot 9H_2O$	70: 20 sup P: 10 PVDF	NaClO$_4$, PC	2.0–4.0	3.3	56	10	[43]
		NaClO$_4$, EC: DMC		3.25, 3.65	54		
		NaPF$_6$, EC: DMC		–	–		
$K_{0.1}Cu^{2+}[Fe^{III}(CN)_6]_{0.7} \cdot 3.5H_2O @ K_{0.1}Ni^{2+}[Fe^{III}(CN)_6]_{0.7} \cdot 4.4H_2O$	75: 20 AB: 5 PTFE	NaClO$_4$, PC	2.2–4.0	ca. 3.25	ca. 75 / 40	10 / 600	[57]
$Na_{0.39}Fe_{0.77}^{3+}Ni_{0.23}^{2+}[Fe^{II}(CN)_6]_{0.79} \cdot 3.45H_2O$	70: 20 AB: 10 PTFE	NaPF$_6$, EC: DEC	2.0–4.0	2.89, 3.35	106	10	[54]
(b) Berlin Green							
$Fe^{3+}Fe^{III}(CN)_6 \cdot 4H_2O$	70: 15 KB: 5 g: 10 PTFE	NaPF$_6$, EC: DEC	2.0–4.0	2.78, 3.47	117	60	[59]
$Fe[Fe(CN)_6]_{1-x} \cdot yH_2O$	80: 10 C-65: 10 PVDF	NaPF$_6$, DMC: DEC	1.0–3.8	–	–	–	[60]

(continued)

Table 4.3 (continued)

Active material	Electrode composition (wt%) (a.m.: C: binder)[a]	Electrolyte[b] (1M salt, solvent mixture)	ΔV (V vs. Na$^+$/Na)	V_{disch} (V vs. Na$^+$/Na)	Q_{disch} (1st cycle) (mAh·g^{-1})	C-rate (mA·g^{-1})	References
[100]-K$_{0.33}$Fe[Fe(CN)$_6$]	80: 10 AB: 10 PTFE	NaClO$_4$, EC: DEC	2.0–3.8	2.8, 3.42	119	0.5 C	[61]
[100]-K$_{0.33}$Fe[Fe(CN)$_6$]@RGO				ca. 3.0	161		
Eu^{3+}+Fe^{III}(CN)$_6$·4H$_2$O	75: 20 AB: 5 PTFE	NaClO$_4$, PC	2.0–4.0	3.33	61	12	[62]
Fe[Co(CN)$_6$]	80: 10 C-65: 10 PVDF	NaPF$_6$, DMC: DEC	1.0–4.0	–	–	–	[60]
(c) Prussian White							
Na$_{1.72}$**Mn**Fe(CN)$_6$	70: 20 AB: 10 PTFE	NaPF$_6$, EC: DEC	2.0–4.2	3.27, 3.57	143	6	[63]
Na$_{1.40}$MnFe(CN)$_6$					134		
Na$_{1.32}$Mn[Fe(CN)$_6$]$_{0.83}$·3.5H$_2$O–ITO	Thin film	NaClO$_4$, PC	2.0–4.0	3.2, 3.6	109	0.5C	[78, 79]
M-Na$_{2-\delta}$Mn[Fe(CN)$_6$]·2H$_2$O	70: 20 KB: 10 PTFE	NaClO$_4$, EC: DEC^g	2.0–4.2	3.17, 3.49	135	15	[64]
R-Na$_{2-\delta}$Mn[Fe(CN)$_6$]				3.44	150		
K$_{1.67}$MnHCF (KNMF-0)	70: 20 sup P: 10 PVDF	NaClO$_4$, EC: DEC	2.0–4.2	3.40, 3.52	40 (100 / 5th cycle)	40	[75]

(continued)

Table 4.3 (continued)

Active material	Electrode composition (wt%) (a.m.: C: binder)[a]	Electrolyte[b] (1M salt, solvent mixture)	ΔV (V vs. Na$^+$/Na)	V_{disch} (V vs. Na$^+$/Na)	Q_{disch} (1st cycle) (mAh·g^{-1})	C-rate (mA·g^{-1})	References
K$_{1.67}$Na$_{0.02}$MnHCF (KNMF-1)				ca. 3.45, 3.54	138		
K$_{1.64}$Na$_{0.21}$MnHCF (KNMF-2)				ca. 3.45, 3.56	138		
K$_{1.59}$Na$_{0.25}$MnHCF (KNMF-3)				3.50, 3.58	118		
Na$_{0.83}$**Fe**[Fe(CN)$_6$]$_{0.97}$ (NaFe0.83)	70: 20 KB: 10 PVDF	NaPF$_6$, EC: DEC	2.0–4.2	3.0	ca. 145	25	[66]
Na$_{1.03}$Fe[Fe(CN)$_6$]$_{0.95}$ (NaFe1.03)					ca. 155		
Na$_{1.24}$Fe[Fe(CN)$_6$]$_{0.94}$ (NaFe1.24)					ca. 135		
Na$_{1.63}$Fe[Fe(CN)$_6$]$_{0.89}$ (NaFe1.63)					ca. 150		
Na$_{0.70}$Fe[Fe(CN)$_6$] (NFF-1)	70: 20 sup P: 10 PVDF	NaClO$_4$, EC: PC	2.0–4.2	2.8, 3.4	107	200	[67]
Na$_{0.90}$Fe[Fe(CN)$_6$] (NFF-2)					113		
Na$_{1.70}$Fe[Fe(CN)$_6$] (NFF-3)				2.8, 3.4, 4.0	121		
Na$_{1.92}$Fe[Fe(CN)$_6$] (NFF-4)					130		
Na$_{1.92}$Fe[Fe(CN)$_6$] (R-FeHCF)	86: 7 KB: 7 PTFE	NaPF$_6$, EC: DEC^g	2.0–4.2	3.0, 3.29	160	10	[68]
Na$_{1.26}$FeFe(CN)$_6$ · 3.8H$_2$O (PB-1)	70: 20 CB: 10 CMC	NaClO$_4$, EC: DEC	2.0–4.0	2.92, 3.42	86.5	20	[69]
Na$_{1.33}$FeFe(CN)$_6$ · 3.5H$_2$O (PB-3)				2.92, 3.04, 3.3	90.8		
Na$_{1.56}$FeFe(CN)$_6$ · 3.1H$_2$O (PB-5)				2.92, 3.04, 3.3	103.6		

(continued)

Table 4.3 (continued)

Active material	Electrode composition (wt%) (a.m.: C: binder)[a]	Electrolyte[b] (1M salt, solvent mixture)	ΔV (V vs. Na+/Na)	V_{disch} (V vs. Na+/Na)	Q_{disch} (1st cycle) (mAh·g−1)	C-rate (mA·g−1)	References
$Na_{1.59}Fe[Fe(CN)_6]_{0.95}·\square_{0.05}$	70: 10 sup P: 10 KB: 10 PVDF	$NaPF_6$, EC: EMC: FEC[h]	2.0–4.0	2.8, 3.3	135 (55[i])/129 (22[i])/99 (−10[i])	100	[70]
PB-H1 (microcubes)	70: 20 C: 10 Na-alg	$NaClO_4$, EC: PC: FEC[g]	2.0–4.2	2.9, 3.4	90	200	[71]
PB-H2 (concave centre)					107		
PB-H3 (spherical-like)					81		
$Na_{1.70}Fe_{2.15}(CN)_6·0.19H_2O$ (Na–PW)	80: 10 C–65: 10 PVDF	$NaPF_6$, EC: PC: FEC[h]	2.25–4.25	2.8, 3.3	145	85	[73]
$K_{1.59}Fe_{2.20}(CN)_6·0.26H_2O$ (K–PW)				2.85, 3.65	150	78	
$K_{0.95}FeHCF$ (NKFF-1)	70: 20 AB: 10 PVDF	$NaClO_4$, EC: DMC	2.0–4.2	2.77, 3.43	140	14	[74]
$Na_{0.08}K_{0.69}FeHCF$ (NKFF-2)							
$Na_{0.12}K_{0.48}FeHCF$ (NKFF-3)				2.55, 3.43			
$Na_{0.17}K_{0.35}FeHCF$ (NKFF-4)							
$Na_{0.46}FeHCF$ (NKFF-5)				2.5, 3.2			
$Na_{1.6}Co[Fe(CN)_6]_{0.90}·2.9H_2O–ITO$	Thin film	$NaClO_4$, PC	2.0–4.0	3.4, 3.8	135	0.6 C	[77, 78]
$Na_{1.95}Co[Fe(CN)_6]_{0.99}·\square_{0.01}·1.9H_2O$	70: 20 g: 10 PTFE	$NaClO_4$, EC: DEC	2.0–4.1	3.2, 3.8	153 / 128	10 / 100	[76]

(continued)

Table 4.3 (continued)

Active material	Electrode composition (wt%) (a.m.: C: binder)[a]	Electrolyte[b] (1M salt, solvent mixture)	ΔV (V vs. Na$^+$/Na)	V$_{disch}$ (V vs. Na$^+$/Na)	Q$_{disch}$ (1st cycle) (mAh·g^{-1})	C-rate (mA·g^{-1})	References
Na$_{0.96}$NiHCF[j]	60: 30 KB: 10 PTFE	NaClO$_4$, EC: DMC: FEC[g]	2.0–4.0	3.22	66	100	[72]
Na$_{1.11}$NiHCF-etch				3.3	90		
Na$_{1.76}$Ni$_{0.12}$**Mn**$_{0.88}$[Fe(CN)$_6$]$_{0.98}$· □$_{0.04}$	70: 20 sup P: 10 PVDF	NaClO$_4$, EC: DMC	2.0–4.0	ca. 3.1	123.3	10	[80]
Na$_{1.67}$**Ni**$_{0.39}$**Co**$_{0.61}$Fe(CN)$_6$	70: 20 AB: 10 PVDF	NaPF$_6$, EC: DEC	2.0–4.2	3.2	92	50	[81]
Na$_{1.96}$**Mn**[**Mn**(CN)$_6$]$_{0.99}$· □$_{0.01}$·2H$_2$O	80: 13 sup P: 7 PVDF	NaClO$_4$, PC	1.3–4.0	1.8, 2.65, 3.55	209	40	[82]

[a] a.m. = active material, C = conductive carbon (AB = acetylene black, sup P = super P, CB = carbon black, KB = ketjen black, C-65 = super C-65, C = conductive carbon, g = graphite); binder (PTFE = polytetrafluoroethylene, PVDF = polyvinylidene fluoride, CMC = carboximethyl cellulose, Na-alg = sodium alginate)

[b] EC = ethylene carbonate, DEC = diethyl carbonate, PC = propylene carbonate, DMC = dimethyl carbonate

[c] In this set of RGOPCs, the formula of Prussian Blue is Na$_{0.81}$Fe^{3+}[Fe^{II}(CN)$_6$]$_{0.79}$·□$_{0.21}$

[d] FeHCF = Na$_{1.1}$Fe[Fe(CN)$_6$]$_{0.9}$·□$_{0.1}$·3.8H$_2$O

[e] PB@C stands for Na$_{0.65}$Fe^{3+}[Fe^{II}(CN)$_6$]$_{0.9}$·□$_{0.07}$·3.45H$_2$O@C

[f] OPB1 = Na$_{0.4}$Fe^{3+}[Fe^{II}(CN)$_6$]$_{0.82}$·□$_{0.18}$·2.75H$_2$O, OPB2 = Na$_{0.39}$Fe^{3+}[Fe^{II}(CN)$_6$]$_{0.82}$·□$_{0.18}$·2.35H$_2$O

[g] 5% FEC (fluoroethylene carbonate)

[h] 2% FEC was added

[i] stands for °C

[j] Na$_{0.96}$NiHCF = Na$_{0.96}$Ni[Fe(CN)$_6$] · 1.02H$_2$O, Na$_{1.11}$NiHCF-etch = Na$_{1.11}$Ni[Fe(CN)$_6$] · 0.71H$_2$O

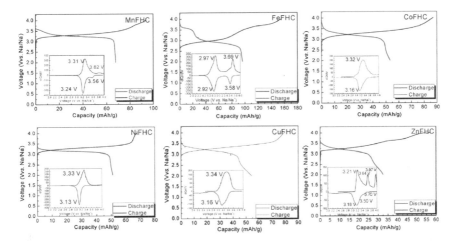

Fig. 4.9 Third charge/discharge curves of Prussian Blue Analogues, KMFe(CN)$_6$ (M = Mn, Fe, Co, Ni, Cu, Zn), against metallic sodium at a current of C/20. Insets show their corresponding chronoamperogram. Reproduced from Ref. [42] with permission of The Royal Society of Chemistry

4.2.1   Prussian Blue Phases

The first work proposing the utilization of Prussian Blue and its analogues as cathode materials for non-aqueous NIB was reported by Goodenough's group in 2012 [42]. For this purpose, several metal hexacyanoferrates, KM^{2+}Fe^{III}(CN)$_6$ (M = Mn, Fe, Co, Ni, Cu, Zn), were prepared by the classical co-precipitation method. All the compounds crystallized in cubic symmetry (*Fm-3m*), except the Zn derivative that does not have the typical Prussian Blue structure and adopts a rhombohedral lattice (*R-3c*), as advanced in Chap. 2.

Galvanostatic charge/discharge curves reflected the different response of these materials as a function of the metal 'M', not only in the voltage of the plateau but also in the capacity values (see Table 4.3 and Fig. 4.9 for further details). The Zn and the Fe complexes exhibited the lowest (35 mAh·g^{-1}) and highest reversible capacities (100 mAh·g^{-1}), respectively. Given its promising behaviour, the Fe complex was evaluated in greater depth, showing a 99% capacity retention after 30 cycles but a poor coulombic efficiency of 80%. As its CV illustrates (inset, Fig. 4.9- Fe-HCF), two plateaus were distinguished at 2.92 and 3.58 V versus Na$^+$/Na. Taking into account that electron filling of C–Fe^{III} (LS d^5) is more energetically favoured than that of N–Fe^{3+} (HS d^5), the high voltage redox process was assigned to the reduction of C–Fe^{III}/Fe^{II}, and the low voltage process to the reaction N–Fe^{3+}/Fe^{2+}. An assignment that is consistent with that already made for non-aqueous LIB[5]. Regardless the capacity values observed, the reversible insertion of Na$^+$ ions in PBA in organic electrolyte was proved, enabling the incorporation of this family of compounds as competent cathode candidates for NIB and so opening a new field of research.

The Zn–Fe System

Lee et al. improved the performance of the Zn–HCF, $Na_2Zn_3[Fe^{III}(CN)_6]_2 \cdot 9H_2O$, reporting reversible capacities close to the theoretical value (60 mAh·g^{-1}) [43]. Even though the reversible capacity value does not sound impressive, it is important to realize that Zn^{2+} is electrochemically inactive and only the redox pair $Fe^{III/II}$ contributes to it. The effect of the electrolyte components was also evaluated within this work. Among the various electrolytes tested, Zn–HCF provided the best electrochemical performance when cycled with $NaClO_4$ in EC: DMC, thus pointing out once again the influence of the solvents and salts in the ionic diffusion at the electrode–electrolyte interface.

The Ti–Fe System

Titanium PBA synthesized at different temperatures (60 and 80 °C), $Na_{0.66}Ti^{3+}[Fe^{II}(CN)_6]_{0.92} \cdot \square_{0.08}$ (Ti–HCF-60) and $Na_{0.7}Ti^{3+}[Fe^{II}(CN)_6]_{0.9} \cdot \square_{0.1}$ (Ti–HCF-80), has also been inspected as possible cathode material for NIB [44]. In both complexes, a pair of redox peaks was differentiated in the charge/discharge curves at 3.0/2.7 and 3.4/3.3 V versus Na^+/Na, which the authors associated to oxidation/reduction of Ti^{4+}/Ti^{3+} and LS $[Fe^{III}(CN)_6]^{3-}/[Fe^{II}(CN)_6]^{4-}$. Ti–HCF-60 provided 92 mAh·g^{-1}, though the capacity drastically diminished along the following cycles. Despite very small compositional differences, lower reversible capacities and higher polarization were observed for the Ti–HCF-80 complex, which possibly result from its higher charge transfer resistance (as indicated the EIS) and the impurities generated by the partial hydrolysis of Ti^{3+} at 80 °C.

The Fe–Fe System: High-Capacity Cathodes

The performance of commercial Prussian Blue, $K_xFe_y^{3+}[Fe^{II}(CN)_6]_z \cdot nH_2O$, versus Na^+/Na was evaluated by Minowa et al. [45] Reversible capacities on the order of 80 mAh·g^{-1} were obtained at 0.125 mA·cm^{-2} along the first discharge, followed however by a fast decay in subsequent cycles. Replacing the conductive carbon acetylene black by ketjen black, the capacity raised by ca. 10 mAh·g^{-1} and its retention improved greatly, given the higher conductivity of the latter; reminding us that every single component of the cell plays a role and matters. Further improvement in the capacity (ca. 20 mAh·g^{-1} more) was achieved by removing water from the Prussian Blue structure by thermal treatment, in an analogous way to what is observed in LIB [1].

Insoluble Prussian Blue, $Fe_4^{3+}[Fe^{II}(CN)_6]_3$, was also investigated as cathode material for NIB [46]. It showed reversible capacities of 141 mAh·g^{-1} at a current density of 50 mA·g^{-1} and two discharge plateaus at ca. 2.8 and 3.2 V versus Na^+/Na. However, the plateau at 3.2 V disappeared along the cycles, supposedly by irreversible insertion of ClO_4^- during the first charge. Let us remember that anion insertion was already observed for *insoluble* PB in LIB [23]. Energy calculations proved that Na^+ ions are reversibly inserted/extracted in the Prussian Blue structure without replacing Fe cations (Fe^{II}, Fe^{3+}) and confirmed that the higher the amount of Na^+ intercalated, the more negative the formation energy becomes and the structure is more stable [46].

Tailoring Prussian Blue: Chemical Etching for High Rate Porous Materials.
Since the abundance of pores in a material can effectively increase its electrochemically active surface area, porous submicron-size cubes of Prussian Blue were produced and tested as cathodes in NIB. First, $Na_{0.4}Fe^{3+}\left[Fe^{II}(CN)_6\right]_{0.82}\cdot\square_{0.18}\cdot2.75H_2O$ (OPB1) was synthesized by the slow growth of Prussian Blue in the presence of PVP (polyvinylpyrrolidone). Next, porous Prussian Blue, $Na_{0.39}Fe^{3+}\left[Fe^{II}(CN)_6\right]_{0.82}\square_{0.18}\cdot$ $2.35H_2O$ (OPB2), was obtained by chemical corrosion of OPB1 with HCl [47]. When cycled versus Na^+/Na, both materials featured initial discharge capacities of ca. 115 $mAh\cdot g^{-1}$ at $50\,mA\cdot g^{-1}$ and better capacity retention than a regular Prussian Blue prepared for comparison. Respectively, OPB1 and OPB2 maintained specific capacities of 93 and 106 $mAh\cdot g^{-1}$ after 50 cycles. The superiority of OPB2 over OPB1 was not only evident in the cycle stability but also in the C-rate capability test. Apparently, the pores of OPB2 are responsible for such improvements, since they shorten the insertion/extraction path lengths, favouring the transport of Na^+ and e^-.

Within this line, a mesoporous PBA was envisaged too [48]. Through a template-free methodology, the surface morphology and porosity of Ni-HCF were adjusted by ageing the synthesis solution in H_2O/DMF (dimethylformamide), simply using different reaction times. Obviously, the crystallite size (11–26 nm) and porosity augmented with the reaction time. Again, in this PBA, all the samples displayed similar capacities (~65 $mAh\cdot g^{-1}$ at $10\,mA\cdot g^{-1}$). However, the rate performance ameliorated the larger their pore size, as higher pore diameters enable facile Na^+ transport.

Consequently, it has been proved that porosity is another important factor to consider in electrode materials, especially for achieving high rate performance.

Suppressed Vacancies and Reduced Water Content
To overcome the main drawbacks of Prussian Blue, poor cyclability and low coulombic efficiency, Guo's group developed a new synthetic method to elude the large amounts of vacancies obtained through the classical co-precipitation preparation [49]. Basically, they mixed $Na\left[Fe(CN)_6\right]\cdot10H_2O$ and HCl in deionized water and stirred the blend at 60 °C for 4 h, obtaining high-quality 300–600 nm in size crystals of Prussian Blue, $Na_{0.61}Fe^{2+}\left[Fe^{III}(CN)_6\right]_{0.94}\cdot\square_{0.06}$ (HQ-Fe–HCF). The electrochemical performance of this HQ-Fe–HCF was contrasted with low-quality Prussian Blue (LQ-Fe–HCF) synthesized through the traditional co-precipitation procedure. Along the discharge profile, multiple plateaus were observed for HQ-Fe–HCF while only two plateaus were differentiated in LQ-FeHCF, which indicates a different Na^+ insertion mechanism for each of them. In addition, HQ-Fe–HCF displayed decreased polarization, improved rate capability and a discharge capacity much higher than LQ-Fe–HCF (Fig. 4.10a). Around 170 $mAh\cdot g^{-1}$, equivalent to a full $2Na^+/2e^-$ insertion/extraction redox process, were reached by HQ-Fe–HCF at $25\,mA\cdot g^{-1}$ without a noticeable capacity loss for 150 cycles, as Fig. 4.10b show. Although it did not reach the target (>99%), the coulombic efficiency raised from 95% in LQ-Fe–HCF to 98% for HQ-Fe–HCF, benefitting from the reduced water content derived from the low amount of vacancies. It was demonstrated therefore that vacancy and water content

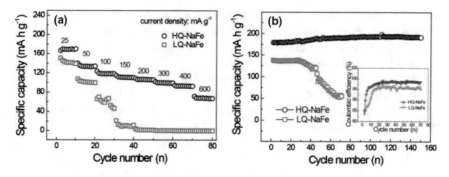

Fig. 4.10 a Rate capability of HQ-NaFePB and LQ-NaFePB and **b** Cycling performance of HQ-NaFePB and LQ-NaFePB under 25 mA·g^{-1}. The inset displays their coulombic efficiencies. Reproduced from Ref. [49] with permission of The Royal Society of Chemistry

strongly affect the performance of PB as cathode materials for NIB, as it had been previously observed in LIB [1, 21].

Inspired by Asakura's [21] and Guo's approaches [49], a highly crystalline Prussian Blue/graphene oxide composite (HC–PB/GO) was synthesized by slow crystallization of PB on GO [50]. As a result of the reaction of Fe_2O_3/GO with $Na_4[Fe(CN)_6]\cdot10H_2O$ and HCl at 60 °C for several days under constant N_2 bubbling, $Na_{0.72}Fe^{3+}[Fe^{II}(CN)_6]_{0.90}\cdot\square_{0.10}\cdot1H_2O$/GO was obtained with a reduced amount of $[Fe(CN)_6]^{4-}$ vacancies and H_2O. The incorporation of GO played a dual role here: to enhance the electronic conductivity of PB and control its particle size (300–500 nm) and morphology (perfect cubes). The reversible capacity achieved by HC–PB/GO for Na^+ storage was larger than that of bare PB (113 vs. 87 mAh·g^{-1} in the first cycle, and 150 vs. 115 mAh·g^{-1} in the second cycle at a current density of 25 mA·g^{-1}). Besides, lower overpotential and an improved rate performance, 107 mAh·g^{-1} at 2000 mA·g^{-1}, were also observed for HC–PB/GO. All this is justified by the higher electronic conductivity and the smaller charge transfer resistance (R_{ct}) resultant of using GO layers, whose interconnecting character reinforced the structural integrity of the material as well. Additionally, the low water and vacancies content of PB along with the presence of GO positively affected the Na^+ diffusion coefficient (D_{Na}^+), obtaining a D_{Na}^+ double than that of bare PB and, thus contributing to improve the kinetics of the material.

Encapsulation of Fe–HCF nanospheres (NSs) in a conductive matrix of graphene rolls (GRs) forming graphene-roll-wrapped PB (Fe-HCF NSs@GRs) was also pursued to avoid side reactions of PBA with the electrolyte (Fig. 4.11a) [51]. One of the advantages this material offers is its direct utilization as a binder-free cathode, decreasing so the weight of inert components. Other relevant aspects are related to its electrochemical behaviour. When Fe-HCF NSs@GRs was tested against metallic sodium, reversible capacities of 110 mAh·g^{-1} were delivered at the moderate-high current density of 150 mA·g^{-1}. Moreover, an improved rate capability (95 mAh·g^{-1} at 1500 mA·g^{-1}) and capacity retention (90% after 500 cycles at 150 mA·g^{-1}) were

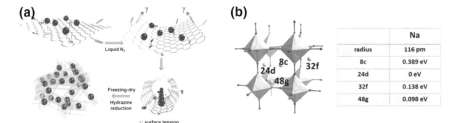

Fig. 4.11 a Formation mechanism of Fe–HCF NSs@GRs 1D tubular hierarchical structure. Reprinted with permission from Ref. [51]. Copyright © 2017, American Chemical Society. **b** Possible Na-ion accommodation sites and their relative binding energies obtained by DFT calculations. Notice that 32f and 48 g could be at any position nearby 8c, but off-centred along the <100> or the <111> directions. In the image, they are shown very far from 8c for clarity. Reproduced from Ref. [62] with permission of The Royal Chemical Society

appreciated over graphene-free nanospheres, evidencing again the beneficial role of conductive graphene.

Enlarging the Capacity of the Fe–Fe System: Further Activation of the Fe–C Redox Pair

As it was referred previously, the upper voltage plateau of Prussian Blue (associated to the redox pair Fe^{III}/Fe^{II}–C) only contributes to a minority of its total capacity [4], so if full participation was achieved, values close to the theoretical capacity could be reached. Using first principle calculations, it was found that the activity of LS Fe–C is highly affected by vacancies and it stabilizes in absence of coordinated water [52]. To investigate the effect that coordinated water exerts in this upper voltage plateau during the insertion/extraction of Na^+, reduced graphene oxide/PB composites (RGOPCs) containing different amounts of RGO were prepared [53]. According to the authors of the original work, the thermal treatment used to reduce graphene oxide would effectively push out the coordinated water, letting a larger number of Fe^{III}–C available to participate in the redox reaction. Essentially, the higher the RGO/PB ratio, the larger the capacity exhibited experimentally (see Table 4.3) and the greater the cycle stability and rate performance. For instance, the composite with the bigger amount of RGO (6.2%) displayed 163 mAh·g^{-1} at 30 mA·g^{-1} and roughly 150 mAh·g^{-1} at 200 mA·g^{-1}, of which 89% was preserved elapsed 500 cycles.

Similarly, a $Na_{0.65}Fe^{3+}[Fe^{II}(CN)_6]_{0.93} \cdot \square_{0.07} \cdot 3.45H_2O$@carbon composite (PB@C) was able to attain larger capacities (130 mAh·g^{-1}) versus Na^+/Na than bare PB (90 mAh·g^{-1}) [52]. As with RGO, the growth of PB nanocubes on carbon chains reduced the H_2O content, so diminishing the number of $[Fe(CN)_6]^{4-}$ vacancies (or $\square$) and increasing the number of Fe–C redox couples available. Furthermore, the composite retained up to 90 mAh·g^{-1} (90% of the initial capacity) even after 2000 cycles at the ultrahigh current density of 20 C (2000 mA·g^{-1}). The low vacancy content, small size and shorter diffusion length between PB and the carbon matrix were considered responsible for such capacity boost and fast reaction kinetics.

Another strategy designed to enlarge the capacity contribution of the LS $[Fe^{III}(CN)_6]^{3-}/[Fe^{II}(CN)_6]^{4-}$ redox pair in Prussian Blue was reported by Yu and coworkers [54], who examined the performance of $Na_{0.39}Fe^{3+}_{0.77}Ni^{2+}_{0.23}[Fe^{II}(CN)_6]_{0.79}\cdot$ $3.45H_2O$. The incorporation of a small amount of inactive nickel into the structure activates the redox couple $[Fe^{III}(CN)_6]^{3-}/[Fe^{II}(CN)_6]^{4-}$ leading to a higher capacity release thereof (ca. 48 mAh·g^{-1}), and thus to a larger total capacity (106 mAh·g^{-1}). According to a recent study [55], doping Fe^{3+}–N with Ni^{2+} would decrease the force constant of the bond $Fe–C\equiv N$ and weaken its crystal field splitting, thus facilitating the transference of electrons from the e_g to t_{2g} orbitals during charge. In addition to activating this redox process, a significantly enhanced cyclability was observed in $Na_{0.39}Fe^{3+}_{0.77}Ni^{2+}_{0.23}[Fe^{II}(CN)_6]_{0.79}\cdot3.45H_2O$ compared to Fe–HCF, which showed a 96% capacity retention and over 97% coulombic efficiency after 100 cycles [54]. Although electrochemically inactive, Ni^{2+} safeguards the structure integrity of the material contributing to extend its cyclability, as we will see now.

Enhancing the Cyclability: The Ni–Fe System and core@shell Structures
Following the concept applied for LIB of using electrochemically inactive metals to stabilize the PB framework [18, 19], the introduction of inactive Ni^{2+} was conceived to eliminate the capacity fading derived from structural changes observed in other M–HCF. In this context, $K_{0.09}Ni[Fe(CN)_6]_{0.71}\cdot6H_2O$ was prepared and electro-chemically tested versus Na^+/Na [56]. A single plateau, associated to the reversible reaction of Fe^{III}/Fe^{II}, was observed at an average voltage of 3 V; and a discharge capacity of 66 mAh·g^{-1} was measured at a current density of 20 mA·g^{-1}. The most relevant feature of this material is by far its cycle stability, as it was capable of retaining 99.7% of its initial capacity after 200 cycles. Such stability is attributed to the absence of structural phase transition, since the structure remains cubic along the Na^+ extraction/insertion process with negligible changes in the lattice parameter (0.29%). For this reason, Ni–HCF is classified within the zero-strain materials for NIB. EIS analysis also revealed a slight increase in the charge transfer resistance of Na^+ diffusion and an almost insignificant change in the apparent D^+_{Na} during cycling, which suggests that the minimal volume change suffered by Ni–HCF not only helps with the structural stability but also preserves stable transport pathways for Na^+.

Interestingly, the core@shell strategy used by Okubo et al. to control the capac-ity and surface stability of PBA in LIB was also useful in NIB [57]. A single cathodic/anodic peak was observed at 3.26 V versus Na^+/Na in the dQ/dV plot of $K_{0.1}Cu^{2+}[Fe^{III}(CN)_6]_{0.7}\cdot3.5H_2O@K_{0.1}Ni^{2+}[Fe^{III}(CN)_6]_{0.7}\cdot4.4H_2O$, manifesting the reversibility of the process. The material reached ca. 75 mAh·g^{-1} at 10 mA·g^{-1}, associated to the reduction of Fe^{III}/Fe^{II} (equivalent to the insertion of 0.7 Na^+) and Cu^{2+}/Cu^+ (corresponding to 0.1 Na^+) as confirmed by XANES, and exhibited an enhanced rate capability compared to Cu–HCF. Ostensibly, the Ni–HCF shell sup-presses the over-sodiation suffered by bare Cu–HCF and the insulating character thereof, in an analogous way to how it does in LIB.

The Role of the Electrolyte and Binder

As it has been gathered in the previous pages, the electrochemical properties of PBA are closely related to the transition metals that compose them. Although it is generally assumed that the electrode material is mainly responsible for the electrochemical behaviour, the electrolyte and binder also influence the performance of a battery. Thus, certain variations observed in the results reported by different authors could be due to differences in the binder or electrolyte employed. To fill in this gap, Piernas et al. conducted a systematic study to find the optimal electrolyte and binder for PB produced by co-precipitation [58]. Among the tested binders (PVDF, CMC, PTFE and EPDM) and electrolytes ($NaClO_4$ or $NaPF_6$ salts in carbonated mixtures –EC: PC, EC: DMC or EC: DEC–, with and without a small amount of FEC as additive), 1M $NaPF_6$ in EC: PC: FEC 49:49:2% vol. and PVDF turned out to be the most suitable electrolyte and binder, respectively. For such combination, Na–Fe–HCF exhibited one of the highest reversible capacities ($130 \, \text{mAh·g}^{-1}$) among those tested, with 99.5% coulombic efficiency and 87% capacity retention after the 40 cycles of the C-rate capability test. Therefore, with the aim of homogenizing the results and facilitating data comparison among different authors, these could be established as standards to be used in future works with Prussian Blue and related compounds (PBA, PW, BG).

4.2.2 Berlin Green Phases

The direct utilization of Berlin Green as a cathode in NIB has been more scarcely reported, although some authors find certain advantages on performing their studies starting from the fully desodiated material.

In 2013, intending to overcome the low coulombic efficiency and poor cycle stability associated to the structural imperfection of the PBA lattice, single-crystal cubic NPs of $FeFe(CN)_6 4H_2O$ were appraised for Na^+ storage [59]. For that purpose, the typical co-precipitation synthesis was followed by heating the mixture at 60 °C for 6 h to allow the formation of the single crystals. As usual for iron-based PB, two plateaus were distinguished at 3.47 and 2.78 V versus Na^+/Na in the CV. After the first cycle, a reversible capacity of $120 \, \text{mAh·g}^{-1}$ (equivalent to $1.52 \, Na^+$) was released at a current density of $60 \, \text{mA·g}^{-1}$, with coulombic efficiencies close to 100%. Besides, an outstanding 95.8% of the initial specific capacity was preserved after 150 cycles. According to the *ex situ* XRD analysis, the almost vacancy-free $FeFe(CN)_6 \cdot 4H_2O$ single crystals remained practically intact during Na^+ insertion/extraction, with only a slight variation in the lattice parameter of 0.23 Å (ca. 2.3%). Logically, its good cyclability was correlated with its structural stability, pointing out once again the importance of obtaining high purity, crystalline and vacancy-free PB-related materials.

Other studies focused their attention and efforts to examine the electrochemical mechanism associated to the Na^+ reversible intercalation in BGA. Pramudita et al. investigated the Na^+ insertion/extraction mechanism in $Fe[Fe(CN)_6]_{1-x} \cdot yH_2O$ and

Fe[Co(CN)$_6$] by *in situ* synchrotron XRD [60]. Curiously, they found that these sodium-free compounds partially convert into sodium-containing species during cell construction, presumably by spontaneous reaction with the electrolyte, experiencing phase segregation. Fe$\left[$Fe(CN)$_6\right]_{1-x} \cdot y$H$_2$O is transformed into a sodium-containing ca. Na$_{0.5}$Fe[Fe(CN)$_6$] major phase and an electrochemically inactive sodium-free Fe[Fe(CN)$_6$] minor phase. On the other hand, Fe[Co(CN)$_6$] evolves into a 'Na-rich' (Na$_{0.224}$Fe[Co(CN)$_6$]) and a 'Na-poor' phase (Na$_{0.108}$Fe[Co(CN)$_6$]) in almost equal quantities, being both electrochemically active. The influence of discharging Fe[M(CN)$_6$] (M = Fe, Co) as a first step versus the traditional initial charge was also addressed but, as confirmed by the minor changes in the lattice/volume, no further Na$^+$ uptake was observed.

First principle simulations on BG showed that Na$^+$ prefer to be accommodated at a face-centred site (24d) rather than at the body-centred site (8c) and, as one could expect, that the {100} plane is active for Na$^+$ diffusion and adsorption [61]. See Fig. 4.11b for a better understanding of the possible Na$^+$ sites. Based on that, Wang et al. prepared {100} facet-capped K$_{0.33}$FeFe(CN)$_6$ microcubes and {100} facet-capped K$_{0.33}$FeFe(CN)$_6$@RGO, using a CTAB-assisted method. Both materials contained a reduced amount of vacancies and lattice water and exhibited enhanced cycling stability and rate capability compared to regular BG. Especially K$_{0.33}$FeFe(CN)$_6$@RGO showed superior performance, delivering 161 mAh·g^{-1} at 0.5C and 132 mAh·g^{-1} at 10 C, of which 92.2 and 90.1% were retained, respectively, after 1000 cycles and 500 cycles. These enhancements were attributed to the high active surface area with all {100} planes exposed, the improved electronic conductivity due to the coupling with RGO and the suppression of water defects.

The uncommon EuFe(CN)$_6$·4H$_2$O was evaluated for NIB as well [62]. Let us remember that, in contrast to other PBA, Eu–HCF adopts an orthorhombic structure (see Fig. 2.3b and c, Chap. 2). Besides, according to the ICP (Inductively Coupled Plasma) analysis, the material did not contain vacancies. When it was cycled versus Na$^+$/Na, a single plateau at 3.3 V, associated with the redox pair [Fe^{III}(CN)$_6$]$^{3-}$/[Fe^{II}(CN)$_6$]$^{4-}$, enabled the full insertion/extraction of 1Na$^+$. This process occurred via a two-phase mechanism with a structural distortion from orthorhombic EuFe(CN)$_6$ · 4H$_2$O to triclinic NaEuFe(CN)$_6$ · 4H$_2$O. Since the volume change was only 1.4%, such phase separation was ascribed to a long-range cooperative rotation of the hexacyanoferrates, which may originate in the weak Eu–N bond. Despite the 3D cyanide structure is preserved, the rotation of the octahedra in the triclinic phase results in two different crystallographic sites for Fe (Fe1 and Fe2) and two Na sites (Na1 and Na2). As Na1Fe1 and Na2Fe2 were arranged in alternate layers along the c-axis and therefore 2D Na$^+$ diffusion in the ab plane was suggested to take place.

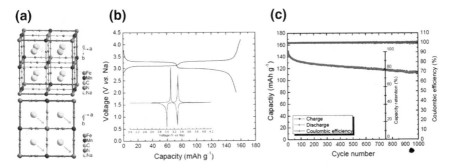

Fig. 4.12 a Crystal structures of cubic $Na_{1.40}MnFe(CN)_6$ (upper) and the alkali-ion displacement along a [111], resulting in rhombohedral symmetry (bottom). Reprinted with permission from Ref. [63]. Copyright © 2003, John Wiley and Sons. **b** Galvanostatic charge and discharge curve of R-Fe–HCF at 10 mA·g^{-1} (chronoamperogram embedded) and **c** Capacity retention of R-Fe–HCF. It was charged at 0.5 C (75 mA·g^{-1}) and discharged at 2 C (300 mA·g^{-1}). Reprinted with permission from Ref. [68]. Copyright © 2015, American Chemical Society

4.2.3   Prussian White Phases

Basically, there are two effective approaches for achieving high energy density materials: (i) increasing their operating voltage by changing or combining different transition metals and (ii) improving their specific capacity by, for example, using alkali-rich compounds. Alkali-rich cathodes, in addition to providing higher capacity, are essential to compensate the lack of sodium that carbon anodes present for full cell operation. In the particular case of PW phases, their higher sodium content also implies less available space for zeolitic water, thus facilitating the access to the 2-e$^-$ redox process and making them attractive candidates as cathode materials for NIB. Certainly, as Fig. 4.8 illustrates, PWA show in general higher capacities than PBA.

The Mn–Fe System
The first work on Prussian White for NIB was reported by Goodenough's group, who contrasted the performance of two sodium-rich Mn–HCF cathode materials, $Na_{1.72}MnFe(CN)_6$ and $Na_{1.40}MnFe(CN)_6$ [63]. It is important to remark that with larger Na$^+$ concentration, a cooperative alkali-ion displacement along the [111] axis takes place (see Fig. 4.12a), lowering the symmetry and inducing a phase transition from cubic (*Fm-3m*) $Na_{1.40}MnFe(CN)_6$ to rhombohedral (*R-3m*) $Na_{1.72}MnFe(CN)_6$, though this phase transition is reversible upon Na$^+$ extraction. When electrochemically tested, both materials exhibited two plateaus upon discharge at 3.27 and 3.57 V versus Na$^+$/Na, associated with the reactions of LS Fe^{III}/Fe^{II} and HS Mn^{3+}/Mn^{2+}, respectively. However, and as expected given its higher initial Na content, the rhombohedral phase delivered larger reversible capacities (143 mAh·g^{-1}) than the cubic phase (134 mAh·g^{-1}). Better cycle stability, retaining 120 mAh·g^{-1} after 30 cycles, and rate capability were observed as well for rhombohedral $Na_{1.72}MnFe(CN)_6$, encouraging further studies based on Na-rich PBA.

The role that interstitial water plays in the electrochemistry of $Na_{2-\delta}Mn[Fe(CN)_6] \cdot zH_2O$ was evaluated by drying a portion of the material at 100 °C in air and the rest at 100 °C under vacuum [64]. As a result, hydrated monoclinic ($P2_1/n$) M-$Na_{2-\delta}Mn[Fe(CN)_6]\cdot2H_2O$ and dehydrated rhombohedral (R-3) R-$Na_{2-\delta}Mn[Fe(CN)_6]$ were respectively formed. The structural differences became soon apparent during the Na^+ insertion/de-insertion process. On Na^+ extraction, both M- and R-$Na_{2-\delta}Mn[Fe(CN)_6] \cdot zH_2O$ were transformed first into a cubic phase (PW → PB process) and finally adopted a tetragonal symmetry (PB → BG). In the subsequent Na^+ intercalation process however, a mixture of M- (hydrated) and R- (dehydrated) was found for M-$Na_2Mn[Fe(CN)_6] \cdot 2H_2O$, while only the R- (phase) was distinguished for R-$Na_2Mn[Fe(CN)_6]$. At 15 mA·g^{-1}, a couple of plateaus at 3.17 and 3.49 V versus Na^+/Na and a reversible capacity of 135 mAh·g^{-1} were noticed for M-$Na_{2-\delta}Mn[Fe(CN)_6] \cdot 2H_2O$, whereas a unique plateau at 3.44 V and a specific capacity of 150 mAh·g^{-1} were exhibited by R-$Na_{2-\delta}Mn[Fe(CN)_6]$. Along with the larger capacity and higher average voltage, the rhombohedral phase showed 75% capacity retention after 500 cycles at 100 mA·g^{-1} with coulombic efficiencies close to 100%. As a result of its promising properties, full cell tests were also built using hard carbon as the anode and the dehydrated R-$Na_{2-\delta}Mn[Fe(CN)_6]$ cathode. Let us clarify that, so far, hard carbons seem the most viable anode for the new NIB technology [65], as graphite is for LIB. Based on the mass of the cathode R-Na_2Mn–HCF, 140 mAh·g^{-1} were delivered at 100 mA·g^{-1} with no obvious capacity decay over 30 cycles, pre-validating the competitiveness of the material in rechargeable non-aqueous NIB.

The Fe–Fe System: Developing New Synthesis Methodologies
At the beginning of 2015, based on their methodology deployed for preparing vacancy-free high-quality PB [49], Guo's group developed a derived procedure for obtaining Na-rich Prussian White [66]. When ascorbic acid and/or protective nitrogen atmosphere are applied, the oxidation of Fe^{2+} and $[Fe^{II}(CN)_6]^{4-}$ can be highly suppressed reaching a notably larger sodium content in PBA/PWA, which evidence the significant influence of the reaction conditions in the sodium quantity present in the final product (see Table 4.3c). Under both conditions, they produced a phase with up to 1.63 Na per f.u., $Na_{1.63}Fe[Fe(CN)_6]_{0.89}$, which exhibited 150 mAh·g^{-1} of reversible capacity, coulombic efficiencies close to 100% after few cycles and an excellent cycle stability of 90% after 200 cycles. The material displayed three plateaus: two small at 3.8 and 3.25 V versus Na^+/Na, and a large one at 2.9 V. As in precedent Na-rich PWA [63, 64], $Na_{1.63}Fe[Fe(CN)_6]_{0.89}$ adopts rhombohedral symmetry and it transforms into cubic during the early stage of charge (or Na^+ extraction), although such phase transition is reverted upon insertion, justifying its cycling stability.

Liu et al. [67] proposed another methodology to synthesize high-quality cubic nanoparticles (NPs) of Prussian White, consisting on the addition of the appropriate amount of sodium citrate to an Fe^{2+} salt and subsequent co-precipitation with hexacyanoferrate. This way, Fe^{2+} ions are first coordinated with citrate forming a chelate complex, which slows down significantly the crystallization speed after the addition of the hexacyanoferrate and yields high-quality materials. With increasing citrate addition, a larger Na content and a lower water quantity were found. When the materials were electrochemically cycled, two reduction peaks were observed at 3.4 and 2.8 V versus Na^+/Na, attributed to LS $[Fe^{III}(CN)_6]^{3-}/[Fe^{II}(CN)_6]^{4-}$ and HS Fe^{3+}/Fe^{2+}. The specific capacity and coulombic efficiency in the first cycle, as well as the rate capability and cycle performance of the prepared materials, were all enhanced the higher the initial amount of sodium. For instance, the Na-rich $Na_{1.70}Fe[Fe(CN)_6]$ exhibited 121 $mAh\cdot g^{-1}$ at 200 $mA\cdot g^{-1}$, retaining 90.9 $mAh\cdot g^{-1}$ after 100 cycles, and around 74 $mAh\cdot g^{-1}$ at 1200 $mA\cdot g^{-1}$. And the sodium poor $Na_{0.70}Fe[Fe(CN)_6]$ delivered ca. 105 $mAh\cdot g^{-1}$ at 200 $mA\cdot g^{-1}$, holding just about 32 $mAh\cdot g^{-1}$ after 100 cycles. By first principles Density Functional Theory (DFT) calculations, the minimum energy of the possible Na^+ sites in $Na_{2-x}FeFe(CN)_6$ (8c, 24d, 32f or 32f') was evaluated too. See Fig. 4.11b for further clarification. These calculations revealed that Na^+ prefers to occupy the 8c (body-centred) metastable site in Na-poor samples, while in Na-rich samples part of Na^+ prefer to occupy the most stable 24d site (face-centred) over the 32f or 32f' sites.

The Fe–PW with the highest sodium occupancy reported so far, $Na_{1.92}Fe[Fe(CN)_6]\cdot 0.08H_2O$ (R-Fe–HCF), is rhombohedral (R-3) and only contains a negligible amount of water [68]. The material was synthesized by adjusting the pH of a $Na_4[Fe(CN)_6]$ solution (0.03M) to 6.5 with ascorbic acid, and maintaining the mixture at 140 °C under stirring in an autoclave for 20 h. During the evaluation of its electrochemical properties for Na^+ storage, two well-defined voltage plateaus at 3.00 and 3.29 V were distinguished on discharge. And roughly 160 $mAh\cdot g^{-1}$ of reversible capacity, equivalent to an energy density of 490 $Wh\cdot kg^{-1}$, were delivered at a current density of 10 $mA\cdot g^{-1}$ (as Fig. 4.12b illustrates). Analogously to other Fe-based PWs, R-Fe–HCF undergoes a phase transition from R-3 to cubic and vice versa respectively upon Na^+ extraction and reinsertion. Nonetheless, the reversibility of the process ensures an excellent cycle stability, and proof of it is that 80% of its initial capacity could be sustained after 750 cycles (Fig. 4.12c). Furthermore, to prove the viability of R-HCF in NIB application, a full cell using hard carbon as anode was assembled. The $r_{N/P}$ (negative to positive capacity ratio) was 1.05. Although the initial coulombic efficiency was low (86%), a discharge capacity of 119.4 $mAh\cdot g^{-1}$ was released in the first cycle at 10 $mA\cdot g^{-1}$, of which 94% was still available after 50 cycles at an average voltage plateau of 3 V.

Using $Na_4Fe(CN)_6$ as precursor and different concentrations of NaCl in presence of HCl, Li et al. prepared a series of 'Na-enriched' $Na_{1+x}FeFe(CN)_6$ samples: $Na_{1.26}FeFe(CN)_6\cdot 3.8H_2O$ (PW-1), $Na_{1.33}FeFe(CN)_6\cdot 3.5H_2O$ (PW-3) and $Na_{1.56}FeFe(CN)_6\cdot 3.1H_2O$ (PW-5) [69]. Curiously, the average particle size of cubic (Fm-3m) $Na_{1+x}FeFe(CN)_6$ decreased with the higher amount of Na^+, while the lattice parameter expanded. As in previous occasions [64, 67], the samples with larger initial sodium content provided higher specific capacities and showed better rate capability. See Table 4.3c for further details. On the other hand, excellent cycling stability was observed in all the samples regardless of the amount of sodium present in them. For example, PW-5 retained 97% of the initial capacity elapsed 400 cycles.

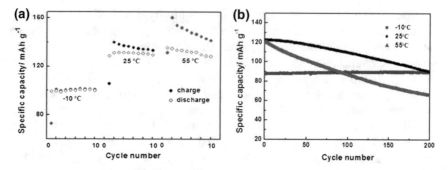

Fig. 4.13 a Specific capacity of $Na_{1.59}Fe[Fe(CN)_6]_{0.95} \cdot \square_{0.05}$ cathode during the first 10 cycles in the 2.0–4.0 V versus Na^+/Na at 10 $mA \cdot g^{-1}$ and **b** Cycle stability of PB at various temperatures in the 2.0–4.0 V versus Na^+/Na at 100 $mA \cdot g^{-1}$. Reprinted from 'Improved cycling performance of Prussian Blue cathode for sodium ion batteries by controlling operation voltage range' from Ref. [70] with permission from Elsevier

Other interesting work evaluated the electrochemical properties of $Na_{1.59}Fe[Fe(CN)_6]_{0.95} \cdot \square_{0.05}$ at various temperatures (−10, 22 and 55 °C) and voltage windows [70]. In the 2.0–4.0 V versus Na^+/Na range, the initial reversible capacities reached 135, 129 and 99 $mAh \cdot g^{-1}$ at 55, 22 and −10 °C, respectively (see Fig. 4.13a). However, the coulombic efficiency and cycle stability followed the inverse trend. Ca. 100% capacity retention was observed at −10 °C after 200 cycles, whereas only 73% and 55% were retained at 22 and 55 °C (Fig. 4.13b). Based on XPS and FTIR data, the capacity degradation was attributed to side reactions with interstitial water close to 4.0 V that would involve the formation of oxygen species causing the rupture of the Fe–N bonds. So, by limiting the voltage window to 2.0–3.8 versus Na^+/Na, those reactions were minimized and the cyclability was highly improved, holding up to 98% of the capacity, for instance, after 300 cycles at 22 °C.

Chemical Etching

In a similar way than for PB, the effect of chemical etching in tailoring Prussian White species was also studied, both by acidic [71] and alkali corrosion [72]. In terms of capacity, no better results were achieved via this methodology. The PW treated with acid, a purely iron-based PW, delivered 110 $mAh \cdot g^{-1}$ at 200 $mA \cdot g^{-1}$. Apparently, acid treatment seemed to induce the exchange of sodium ions by hydrogen, thus lowering the sodium content. As for the species corroded in alkali media, a NiFe-PWA, ca. 90 $mAh \cdot g^{-1}$ were obtained at 100 $mA \cdot g^{-1}$. The cyclability and rate performance, however, benefited from these treatments. PW treated with acid showed a capacity retention of 74% after 500 cycles, and NiFe-PWA held an outstanding 83.2% of the initial capacity after 5000 cycles at 500 $mA \cdot g^{-1}$. Besides, around 80 $mAh \cdot g^{-1}$ were exhibited by acidic-etched PW at the high current density of 5 $A \cdot g^{-1}$, and NiFe-PWA-etched maintained 70.9 $mAh \cdot g^{-1}$ at 44.4 C (ca. 4 $A \cdot g^{-1}$). It is important thus to note here that these results probably represent the best high rate performance reported so far for PBA/BGA/PWA applied to non-aqueous rechargeable NIB.

Enhancing the Energy Density of the Fe–Fe & PBA Systems: Hybrid Batteries
Cubic Na–PW, $Na_{1.70}Fe_{2.15}(CN)_6 \cdot 0.19H_2O$ or $Na_{1.48}Fe[Fe(CN)_6]_{0.87}\square_{0.13} \cdot 0.17H_2O$, and K–PW, $K_{1.59}Fe_{2.20}(CN)_6 \cdot 0.26H_2O$ or $K_{1.33}Fe[Fe(CN)_6]_{0.83}\square_{0.17} \cdot 0.22H_2O$, with low water content, were synthesized following a new mild synthetic procedure that did not require acidic reactants [73]. Obtaining alkali-rich phases with cubic symmetry could be beneficial, since the species formed during the cycling process (PB and BG) are cubic as well and no structural transitions would be required in principle through it. Let us remember, for example, that the phase transition from cubic to rhombohedral has been reported at $x \geq 1.7$ for the sodiated species $Na_xFe[Fe(CN)_6] \cdot xH_2O$ [63]. Conversely, the initial capacity would be limited because cubic phases do not present as high alkali content as the rhombohedral or monoclinic ones. When tested versus Na^+/Na, both A–PW compounds (A = Na, K) displayed two plateaus and delivered reversible capacities above 145 $mAh \cdot g^{-1}$ at 1 C (ca. 80 $mA \cdot g^{-1}$), as Fig. 4.14a and b show. Na-PW featured better high rate capability, whereas K–PW exhibited superior coulombic efficiency and cycling stability, retaining 80% of the initial capacity after 500 cycles (Fig. 4.14c). A stimulating advantage that K–PW showed with respect to Na–PW was the 0.35 V boost in the high voltage redox plateau (see Fig. 4.14 and Table 4.3c). Property that should lead to a profitable increment of the gravimetric energy density in a full cell configuration. Via electrochemical and chemical analyses, this beneficial potential shift of K–PW towards higher values was confirmed to be due to the preferential insertion of K^+ ions instead of Na^+. On the other hand, the low voltage plateau stayed at the same value than that of Na–PW, indicating that only Na^+ insertion occurs in this low voltage redox process. Therefore, it was concluded that the synergistic effects of the hybrid Na^+ and K^+ insertion in the framework of K–PW are responsible for the long cycle stability of the material, thus making K–PW an even more attractive cathode material than Na–PW for real market applications.

To further clarify the role of K^+ ions in alkali iron hexacyanoferrate, two systematic studies were performed on Na_xK_y-Fe–HCF and Na_xK_y-Mn–HCF [74, 75]. In Liao's work, regardless of the sodium to potassium ratio (0:1, 1:1, 3:1, 7:1, 1:0), all the Na_xK_y-Fe–HCF Prussian White NPs showed initial discharge capacities close to 140 $mAh \cdot g^{-1}$ at 14 $mA \cdot g^{-1}$ [74]. As for Liu's study on Na_xK_y-Mn–HCF, the material containing the highest potassium content and still some amount of sodium displayed a better rate performance and cycle stability than the rest [75]. Undoubtedly, the biggest difference among samples was found in the voltage profile, where redox process shifted to higher voltages the higher the potassium content (see Table 4.3) [74, 75]. Observations agree with the results reported by Piernas et al. [73]. The number of redox peaks observed also varied from sample to sample [74]. The splitting of the higher voltage redox peak in Na–Fe–HCF (no potassium) suggested the intercalation of Na^+ in both 8c and 24d sites of the cubic lattice, while the single peak distinguished for Na_xK_y–Fe–HCF (potassium-containing) denoted that only the 8c site is available for the insertion of the bigger K^+. In addition, it was discovered that the volumetric

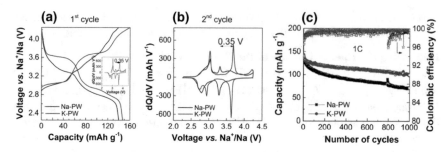

Fig. 4.14 **a** First cycle voltage profile (inset first cycle dQ/dV vs. voltage curve), **b** second cycle dQ/dV versus voltage curve and **c** long-term cyclability (charge/discharge capacities and coulombic efficiencies) of Na–PW (black) and K–PW (violet), when cycling versus Na⁺/Na. Reprinted from 'Higher voltage plateau cubic Prussian White for Na-ion batteries', Ref. [73], with permission from Elsevier

expansion and phase transition (from cubic to rhombohedral) observed upon Na^+ insertion in Na–Fe–HCF are prevented during the dual K^+ and Na^+ insertion process in Na_xK_y–Fe–HCFs, as K^+ ions are too large to move away from the 8c site, enhancing this way their cycling performance.

The Co–Fe System

The production of highly crystalline $Na_{1.95}Co[Fe(CN)_6]_{0.99} \cdot \square_{0.01} \cdot 1.9H_2O$ with suppressed structural defects by citrate-assisted crystallization was reported by Wu and collaborators [76]. Analogously to the Fe–HCF derivative [67], the citrate acts as a chelating agent, slowing down the crystallization speed. When electrochemically tested versus Na^+/Na, CoFe–PWA experiences a 2e- redox process, in contrast to the single-electron process observed in CoFe-PBA for LIB and NIB [8, 42]. Specifically, two voltage plateaus were distinguished at 3.8 and 3.2 V versus Na^+/Na, corresponding to the redox reactions of the Fe^{III}/Fe^{II} and Co^{3+}/Co^{2+}. Clearly, the high sodium content of CoFe–PWA was reflected in the initial discharge capacity (153 mAh·g^{-1}) and the coulombic efficiency (ca. 98%), which could enable an energy density of 510 Wh·kg^{-1}. The material also exhibited good rate capability and long-term cyclability, retaining 90% over 200 cycles at the moderate current density of 100 mA·g^{-1}, even though it experiences a phase transition from rhombohedral to cubic on the charging process and vice versa along discharge.

Thin Films: The Mn–Fe and Co–Fe Systems

Some of the few works involving the use of PW thin films for Na-ion storage were carried out by Moritomo's group. Specifically, they studied the electrochemical properties of $Na_{1.6}Co[Fe(CN)_6]_{0.90} \cdot 2.9H_2O$ [77, 78] and $Na_{1.32}Mn[Fe(CN)_6]_{0.83} \cdot 3.5H_2O$ [78, 79] thin films (of ca. 1.1 μm). In the cobalt complex, a pair of discharge plateaus were differentiated at 3.8 and 3.4 V versus Na^+/Na, accordingly ascribed to the reduction of Fe^{III} and Co^{3+}. For the manganese analogue, these appeared at 3.6 and 3.2 V and were related to the reduction of Mn^{3+} and Fe^{III}. Reversible specific capacities evidencing a two-e$^-$ reaction were observed in Co–HCF (135 mAh·g^{-1} at 0.6 C) and Mn–HCF (109 mAh·g^{-1} at 0.5 C), with coulombic efficiencies close to 97 and

95% and capacity retentions of 90% and 71% after 100 cycles, respectively. Along with the superior cyclability, Co–HCF exhibited a remarkable C-rate performance, which originated in the suppression of the habitual phase transition (cubic $\rightarrow$ rhombohedral). Note the lower Na content compared to the previous CoFe–PWA [76]. Conversely, the Mn complex was highly affected by the Jahn–Teller distortion of Mn^{3+}, as demonstrated as well by its C-rate capability.

Ternary Phases: The Ni–Mn–Fe and Ni–Co–Fe Systems
Given the zero-strain nature and excellent cyclability of $Na_2NiFe(CN)_6$ (Na–Ni–HCF), Ni doping into high-capacity $Na_2MnFe(CN)_6$ (Na–Mn–HCF) and $Na_2CoFe(CN)_6$ (Na–Co–HCF) have resulted in high-performance ternary compounds $Na_{1.76}Ni_{0.12}Mn_{0.88}[Fe(CN)_6]_{0.98} \cdot \square_{0.04}$ (Na–NiMn–HCF) [80] and $Na_{1.67}Ni_{0.39}Co_{0.61}Fe(CN)_6$ (Na–NiCo–HCF) [81]. The electrochemical properties of Na–NiMn–HCF lied between those obtained by Na–Mn–HCF and Na–Ni–HCF, with a voltage plateau at ca. 3.1 V versus Na^+/Na, a first discharge capacity of 123.3 $mAh \cdot g^{-1}$ and a coulombic efficiency of 92.7%. The material maintained almost 84% of the initial capacity after 800 cycles, approaching somehow the superior cycle stability observed for Na–Ni–HCF (99.7% retention after 200 cycles) [56] and suggesting that just a 12% of inert Ni^{2+} in the Mn–HCF framework is enough to break the long-range order of the structural perturbations derived from the redox reaction of Mn^{3+}/Mn^{2+}. On the other hand, Na–Ni–Co–HCF displayed a single plateau at 3.2 V associated only with the reduction of $[Fe^{III}(CN)_6]^{3-}$ [81], despite both metals (Fe and Co) had proved to be actively involved in the binary system CoHCF [76]. Initial discharge capacities (92 $mAh \cdot g^{-1}$) intermediates to the nickel and cobalt complexes were delivered, along with low polarization (0.17 V), good cycle stability (89.5% after 100 cycles), and excellent rate performance (70 $mAh \cdot g^{-1}$ at 800 $mA \cdot g^{-1}$) [81]. Analogous to previous works, both transition metals play a role: Ni^{2+} acts as framework support maintaining structural integrity [56, 57] and both Ni^{2+} and Co^{2+} help to activate the $[Fe^{III}(CN)_6]^{3-}/[Fe^{II}(CN)_6]^{4-}$ couple [54] increasing the specific capacity.

Using commercial hard carbon as the anode, a full cell based on the ternary Na–NiMn–HCF was assembled. It delivered a capacity of 90.8 $mAh \cdot g^{-1}_{cathode}$ under 100 $mA \cdot g^{-1}$ with an initial coulombic efficiency of 88.2% that increased to 95% in subsequent cycles. The energy and power density were estimated at 81.7 $Wh \cdot kg^{-1}$ and 90 $W \cdot kg^{-1}$, respectively. Promisingly, after 200 cycles, the full cell still retained 78.9% of the initial capacity.

The Mn–Mn System
Monoclinic $Na_{1.96}Mn[Mn(CN)_6]_{0.99} \cdot \square_{0.01} \cdot 2H_2O$ (Mn–HCMn) NPs [82], featuring almost full sodium occupancy and no vacancies, were produced via utilization of a large excess of Na^+ [63] during the synthesis. Three plateaus (contributing each one with approx. 70 $mAh \cdot g^{-1}$) appeared during the galvanostatic discharge at 3.55 V ($Mn^{3+}/Mn^{2+}–N\equiv C–Mn^{III}$), 2.65 V ($Mn^{2+}–N\equiv C–Mn^{III}/Mn^{II}$) and 1.8 V versus Na^+/Na ($Mn^{2+}–N\equiv C–Mn^{II}/Mn^{I}$), delivering a total reversible capacity of 209 $mAh \cdot g-1$ (Fig. 4.15a). To date, this is the highest specific capacity reported for any BGA/PBA/PWA cathode. Although it is necessary to clarify that the wider

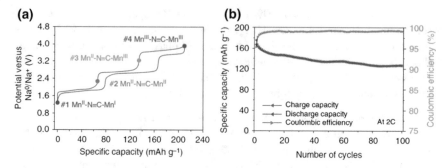

Fig. 4.15 **a** Galvanostatic charge and discharge curve at C/5 of MnHCMn and **b** Cycling stability (specific capacities and coulombic efficiency) of MnHCMn at 2 C. Reprinted by permission from Springer Nature from 'Manganese hexacyanomanganate open framework as a high-capacity positive electrode material for Na-ion batteries', Ref. [82], advance online publication, 14 October 2014

voltage range explored, from approximately 1.3–4.0 V versus Na^+/Na, enables a third Na^+ to be inserted along with a third redox reaction ($Mn^{2+}-N \equiv C-Mn^{II}/Mn^{I}$). This ability of Mn^{II} to remain in the PB framework while is reduced to Mn^{I} seems to be specific of hexacyanomanganates. Let us remember that upon insertion of two or more Li^+ ions in HCF, the crystal structure apparently breaks apart undergoing a conversion reaction,[34] as described in Sect. 4.1.2. Similarly to other PWA, Mn–HCMn experienced phase transitions during Na^+ storage/uptake. During the oxidation reaction, pristine monoclinic ($P2_1/n$) Mn–HCMn transforms to orthorhombic ($P222_1$) and finally to the fully oxidized cubic phase (Fm-$3m$), being these transitions reversed during reduction. The fully reduced sample, however, seems to be a mixture of the pristine monoclinic and a more distorted monoclinic phase.

In summary, the electrochemical performance of PBA/BGA/PWA is closely linked to their crystal structure and composition. As discussed above, the existence of $[Fe^{II}(CN)_6]^{4-}$ vacancies implies a reduction of the Na^+ content in the PB lattice (since less charge has to be balanced), negatively affecting the capacity. In addition, these vacancies translate into more water molecules in the structure, which may block Na^+ insertion and further decrease their capacity and cycle life. Therefore, high-quality PBA/BGA with suppressed vacancy and water contents, and alkali-rich PWA (inherently with low water and $[Fe^{II}(CN)_6]^{4-}$ vacancies content), are highly recommended for NIBs (and non-aqueous battery) applications, as they feature improved capacity and cycle life with respect to regular PBA. The production of PB (or PBA) composites with highly conductive carbon species (reduced graphene oxide, graphene rolls, etc.) also procured better rate capability and higher cycle stability, given the effective suppression of the water content. In addition, we have seen that the incorporation of inactive (or sometimes active) metals enhances too the long-term cyclability and rate performance, and in the particular case of HCF, they help to further activate the Fe–C redox pair. Last but not least, synergistic effects of a dual A^+ insertion into PBA/PWA have also been found to be beneficial for the electrochemistry, not only enlarging the cycle life of the material but also increasing the voltage at which the redox reactions occur.

Further details about the experimental conditions (electrode composition, electrolyte, voltage window, etc.) employed for each system described within this 'NIB' section can be found in Table 4.3. The materials have been organized in the same way than in Tables 4.1 and 4.2.

Aside from the information collected in Chaps. 3 and 4, we would like to mention three recent reviews that may also be of interest to the reader [83–85].

4.3 Prussian Blue and Analogues in Other Non-aqueous Batteries

The recent interest in other chemistries 'beyond Li-ion batteries' expands as well to other insertion cations besides Na-ion, pursuing to achieve higher energy densities than those obtained for NIB and using elements that are also more abundant than Li. Prussian Blue materials have been assessed in other beyond lithium technologies, including monovalent K-ion batteries (KIB), divalent Mg-ion batteries (MIB) and Ca-ion batteries (CIB), and trivalent Al-ion batteries. We will see that K-ion batteries show comparable or even superior performance to Li- and Na-ion, as there is a good size matching between potassium and the zeolitic cavities, while multivalent batteries present more challenges.

4.3.1 Prussian Blue and Other Phases as Cathodes for K-ion Batteries

The appeal of KIB resides in the high natural abundance of potassium as well as in its electrochemical potential (-2.925 V vs. SHE), which is the closest to that of lithium (-3.045 V vs. SHE) among the alkali metals and therefore ensures high energy density [86]. On the other hand, assembling cells with potassium metal entails more difficulties than with lithium or sodium, given its faster reactivity with low concentrations of H_2O. Additionally, the choice of electrolyte is another matter to consider, as the potassium analogous salts may not be as soluble in the solvent mixtures already tested for other technologies.

Prussian Blue
The first KIB utilizing a Prussian Blue cathode in non-aqueous media was conceived by Eftekhari, back in 2004 [87]. This cell consisted of a PB thin film deposited onto a Pt substrate as working electrode, a metallic potassium anode which exerted as counter and reference electrode, and KBF_4 in EC: EMC (7:3) as electrolyte. The CV of this cell displayed a couple of peaks at 0.86 and 0.18 V versus SHE (ca. 3.79 and 3.11 V vs. K^+/K), associated respectively to the oxidation and the

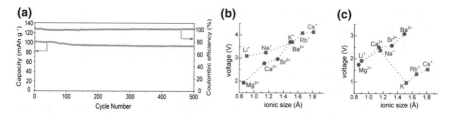

Fig. 4.16 **a** Long-term cyclic performance of Fe–HCF versus K^+/K in 1M KPF_6 in EC: DEC at a rate of 5 C. Reproduced from Ref. [89] with permission of The Royal Society of Chemistry. **b** Calculated voltages for the intercalation of alkali (A) and alkali-earth (AE) ions into Fe–HCF as a function of the ionic radius associated with the redox process. $Fe_{LS}^{III}/Fe_{LS}^{II}$, i.e. from $Fe[Fe(CN)_6]$ to $AFe[Fe(CN)_6]$ or $AE_{0.5}Fe[Fe(CN)_6]$, and **c** $Fe_{LS}^{3+}/Fe_{LS}^{2+}$, i.e. from $AFe[Fe(CN)_6]$ to $A_2Fe[Fe(CN)_6]$ or $AE_{0.5}Fe[Fe(CN)_6]$ to $AEFe[Fe(CN)_6]$. Reprinted with permission from Ref. [93]. Copyright © 2013, American Chemical Society

reduction of PB to BG and PW. However, according to the UV–visible absorption spectroscopy performed on charged and discharged electrodes, exclusively the upper redox plateau (C–Fe^{III}/Fe^{II}) at 3.11 V versus K^+/K seemed to be accessible during the galvanostatic tests. About 78 mAh·g^{-1}, corresponding to 90% of the theoretical capacity, were achieved in the first discharge at C/10 ($\approx$8.7 mA·g^{-1}). Moreover, it exhibited a capacity fade of only 12% after 500 cycles, demonstrating more reversible cyclability for insertion/extraction of K^+ than for Li^+.

Berlin Green

With the recent upsurge in beyond LIB, Zhang et al. resumed the investigations of this material and tested $K_{0.22}Fe[Fe(CN)_6]_{0.805} \cdot \square_{0.195} \cdot 4.01H_2O$ NPs for KIB [88]. As a result of the interstitial water present in the compound under study, the initial coulombic efficiency was poor (44%), but it increased gradually, reaching ca. 95% in the 25th cycle. In the galvanostatic measurements, a single redox process related to the cathodic reaction C–$Fe^{III/II}$ was distinguished at ca. 3.18 V versus K^+/K, and similar discharge capacities (77 mAh·g^{-1}) to that observed by Eftekhari were delivered and sustained for 50 cycles at 50 mA·g^{-1}.

Interestingly, another BG prepared by Shadike et al., $Fe^{3+}Fe^{III}(CN)_6$, exhibited two redox processes, with reductions at 3.4 (Fe^{III}/Fe^{II}–C) and 3.2 V versus K^+/K (Fe^{3+}/Fe^{2+}–N) [89]. As a result, the material released an initial discharge capacity of 121 mAh·g^{-1} at 0.5 C (25 mA·g^{-1}), of which 117 mAh·g^{-1} were retained after 100 cycles. Even at 5 C (625 mA·g^{-1}), $FeFe(CN)_6$ proved a good cycle stability, holding 93% of its initial capacity elapsed 500 cycles (Fig. 4.16a). In addition, it exhibited negligible polarization and coulombic efficiencies close to 100%.

In both studies, ex situ XRD confirmed the retention of the cubic structure at all stages of cycling, unlike what happens with the Na^+ systems when they exceed a certain amount of cation [66, 68]. In fact, a minimal lattice and volume expansion (0.39% and 1.18%) of the cubic unit cell was encountered for the material synthesized by Shadike during the insertion process, so manifesting its zero-strain nature [89]. Such structural stability, probably derived from the low deformation while accommodating the K^+ in the body-centred 8c sites, was also consistent with its outstanding cyclability [89].

Going one step further, Zhang and coworkers assembled a carbon black super P/Fe–HCF NPs full cell that delivered up to 65 mAh·$g_{cathode}^{-1}$ at an average discharge voltage of 2.5 V under a current density of 100 mA·g^{-1}, and retained 64 mAh·g^{-1} even after 50 cycles [88]. This way, the feasibility of PB-based non-aqueous K-ion rocking chair full cells was proved.

Prussian White
The electrochemical properties of several PWA, cubic $K_{2-x}M[Fe(CN)_6]\cdot mH_2O$ (M = Mn, Fe, Co, Ni, Cu), were evaluated as well in non-aqueous KIB [90, 91]. MnFe-PWA and purely iron-based PW displayed two plateaus, which were observed: at 3.7 and 3.56 V versus K^+/K in the former (associated sequentially to the reduction of HS Mn^{3+}/Mn^{2+}–N and LS Fe^{III}/Fe^{II}–C) [91] and at much more differentiated voltages in the latter, 3.9 and 3.2 V versus K^+/K [90]. For the other phases, CoFe-PWA, NiFe-PWA and CuFe-PWA, a single redox process was noted respectively at 3.5, 3.7 and 3.75 V [90]. The absence of the second plateau for CoFe-PWA, commonly observed in NIB [76], was probably due to the higher extraction voltage of K^+ versus Na^+. Consistently with the number of redox active centres, the discharge capacity released by PW (110 mAh·g^{-1}) was higher than their CoFe-PWA (60 mAh·g^{-1}), NiFe-PWA (63.4 mAh·g^{-1}) and CuFe-PWA (35 mAh·g^{-1}) counterparts. As for the long-term cycling, CoFe-PWA experienced the fastest capacity fading, NiFe-PWA and CuFe-PWA presented relatively stable cycle life, while PW showed the best capacity retention (81% after 100 cycles). Specifically for Mn–PW, $K_{1.89}Mn[Fe(CN)_6]_{0.92} \cdot 0.75H_2O$ (Mn–HCF) and sodium-induced $K_{1.70}Mn[Fe(CN)_6]_{0.90} \cdot 1.1H_2O$ (NI–Mn–HCF) were studied [91]. Mn–HCF exhibited 142.4 mAh·g^{-1} at 0.2 C, though its capacity faded along cycling. NI–Mn–HCF, however released 85 mAh·g^{-1} at 1 C that increased over the next 20 cycles, displaying a more stable cycle life. *Ex situ* XRD data detected a lattice expansion upon K^+ extraction on Mn–HCF. A tendency that is surprisingly opposed to that observed in Fe–HCF [88, 89] and that could be explained by the Jahn-Teller effect of Mn^{3+}.

The effect of the crystallite size of Prussian Blue/White was also investigated [92]. Using the citrate chelation route previously applied in NIB [67], K–PW or K–PB samples with small (≈ 20 nm, **S**), intermediate (ca. 170–200 nm, **M**) or large size (micron size, **L**) were synthesized. The reversible capacities at 10 mA·g^{-1} varied from 140 mAh·g^{-1} for K–PW–**S** to 125 mAh·g^{-1} for K–PW–**M** and dropped down to 10 mAh·g^{-1} for K–PB–**L**, manifesting that the smaller the particle size, the higher the capacity delivered. Although, it is necessary to highlight that the composition of sample **L** is very different to the other samples. In this work, the effect of adding FEC to the electrolyte was addressed too. Analogous to what was observed in NIB, higher coulombic efficiency and capacity retention are obtained in the presence of FEC.

Preferred Crystallographic Sites of Monovalent and Divalent Cations in PB
A DFT study performed by Ling et al. focused on the effect that the ionic radii of alkali and alkali-earth metals have on the lattice energies during the intercalation

process in Prussian Blue [93]. The calculations revealed that Li^+, Na^+, Mg^{2+} and Ca^{2+} prefer to intercalate at the face-centred *24d* site while larger cations, such as K^+, Rb^+ and Cs^+, accommodate at the body-centred *8c* site. See Fig. 4.11b for further clarification of the possible A^+ sites. In addition, a correlation between the cation size and their insertion voltage into Fe–HCF was found (Fig. 4.16b). The larger ions insert at higher voltages (3.08 V for Li^+, 3.23 V for Na^+ and 3.70 for K^+), which is consistent with experimental results [73, 74]. For the low voltage redox process (Fig. 4.16c), however, there is a striking drop in the calculated voltage when comparing Na^+ to K^+ insertion. Phenomenon that was explained by the different site that Na^+ and K^+ occupy in the lattice, thus increasing the repulsion among intercalated cations at high concentrations thereof, which contributes to the total energy of the system. Nevertheless, this big drop in voltage is not experimentally observed in PW [73].

4.3.2 Prussian Blue as Cathode for Mg-ion Batteries

The main benefits of using Mg metal anodes are its apparent dendrite-free deposition and its high theoretical volumetric capacity. However, Mg^{2+} ions show sluggish diffusion and the electrolytes developed for MIB are generally known to corrode the current collectors deployed in LIB or NIB [94]. An attempt to overcome these drawbacks and achieve fast ionic transport in the cathode, while exploiting the benefits of a Mg anode, consists of using a Mg–Li dual electrolyte [95]. Within this context, Nazar's group cycled micrometric $Fe[Fe(CN)_6]_{0.95} \cdot nH_2O$ (n = 2.3 or 0.7) versus Mg^{2+}/Mg, using $[PhMgCl]_2 \cdot AlCl_3$ in THF as electrolyte. Since Mg^{2+} did not intercalate into BG, they added Li^+ in the form of LiCl. With increasing LiCl concentration of up to 0.5 M, the electrolyte yielded a higher conductivity and the reversible capacity was enhanced, attaining 125 mAh·g^{-1} at a current density of 10 mA·g^{-1} and an average voltage of 2.3 V versus Mg^{2+}/Mg. All the electrochemical capacity originated solely in the insertion/extraction of Li^+ and no Mg^{2+} was intercalated into PB. Nevertheless, dendrite-free surface typical of Mg electrodeposition was observed on the anode after long-term cycling, which constitutes an advantage over LIB.

4.3.3 Prussian Blue and Analogues as Cathodes for Ca-ion Batteries

Ca^{2+} ion presents a large ionic radius (1.18 Å) [6] and slow diffusion but, as Mg^{2+}, it benefits from a 2-e$^-$ redox process per each cation inserted. Thus, high capacity and volumetric energy density are expected from using metallic Ca anodes. Other additional advantages include its abundance and a standard electrode potential (-2.87 V vs. SHE) [96] larger than magnesium (-2.37 V vs. SHE).

The intercalation of Ca^{2+} in PBA in non-aqueous media was first reported by Padigi et al. [97] To expand the HCF framework, they incorporated Ba^{2+} in the structure yielding $K_2BaFe(CN)_6$. Its electrochemical response in $Ca(ClO_4)_2$/acetonitrile was almost null due to the strong electronic interaction of Ca^{2+} with the host material. However, the addition of 17% water to the electrolyte enabled the generation of a hydration sphere that shielded the coulombic interactions of Ca^{2+} with the host material and resulted in sharp oxidation and reduction peaks at 0.28 and 0.11 V versus Ag/AgCl and a discharge capacity of 62.2 mAh·g^{-1}. Lower and higher amounts of water led to the lack of Ca^{2+} intercalation and the dissolution of the cathodic material, respectively, indicating that a critical number of water molecules are required to form a favourable solvation sphere around the cations to be intercalated. Additional EIS measurements revealed the decrease of the charge transfer resistance of Ca^{2+} at the electrode interface upon water addition. Nevertheless, this would preclude the use of metallic calcium as anode since it reacts with water.

Another challenge to overcome in Ca-ion electrochemistry is to find appropriate electrolytes that enable reversible plating and stripping of Ca [98]. In this regard, Shiga et al. investigated the insertion of Ca^{2+} into $K_{0.1}Mn_1Fe_{1.1}(CN)_6\cdot4H_2O$ with some organic solvents and an ionic liquid, using a Pt counter electrode [99]. Among them, the ionic liquid electrolyte $Ca(CF_3SO_3)$ in DEME$^+$TFSA$^-$ (being in DEME$^+$TFSA$^-$ $= N$, N-diethyl-N-methyl-N(2-methoxyethyl)ammonium bis(trifluoro-methanesulfonyl)amide) showed the best results. Since calcium plating proved to be challenging with such ionic liquid, they replaced Ca by Mg metal and fabricated a hybrid Ca^{2+}/Mg^{2+} battery that released ca. 62 mAh·g^{-1} at 1.55 V during discharge, and 67 mAh·g^{-1} at 2.1 V upon charging in the first cycle. Although with large capacity fading, this hybrid battery was able to charge and discharge for 20 cycles.

Lipson and coworkers also addressed the challenge of inserting Ca^{2+} in $Na_{0.2}Mn[Fe(CN)_6]$ (Mn–HCF) but using a carbon anode and a 0.2 M $Ca(PF_6)_2$ in EC: PC (3:7) electrolyte [100]. In this case, only $Mn^{3+/2+}$ reduction takes place at ca 3.4 V versus Ca^{2+}/Ca during discharge, delivering reversible capacities close to 80 mAh·g^{-1}. Despite the Mn–HCF was evaluated for Ca–IB after electrochemical desodiation of $Na_{1.1}Mn[Fe(CN)_6]$, dual Na^+ and Ca^{2+} insertion/extraction was suggested by EDX and XPS. By coupling the desodiated Mn–HCF cathode with a calciated tin anode, they also built a full Ca-ion battery. Similar capacities to those attained versus the carbon anode and an average voltage of 2.6 V were observed for this full cell, although the capacity decayed approximately 50% after 35 cycles.

With a similar system to that of Padigi, i.e. activated carbon, Ag/AgCl and $Ca(TFSI)_2$ in acetonitrile, two peaks related to the insertion and extraction of Ca^{2+} into $K_xNi[Fe(CN)_6]$ (zero-strain for NIB) were observed at –0.15 and 0.6 V versus Ag$^+$/Ag in the CV [101]. Galvanostatic tests displayed a discharge plateau at 0–0.2 versus Ag$^+$/Ag with reversible capacities up to 50 mAh·g^{-1}, of which ca. 40 mAh·g^{-1} were sustained for at least 12 cycles. Nevertheless, the presence of Ca^{2+} and some K^+ was confirmed by EDX and XPS analyses after the first charge–discharge, revealing again a dual ion transport nature of the battery.

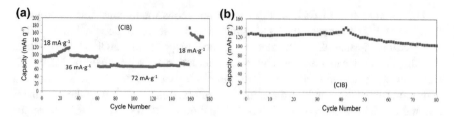

Fig. 4.17 a Reversible specific capacity versus cycle number at three different current densities and **b** reversible specific capacity at 125 mA·g^{-1} of PB in Ca(ClO$_4$)$_2$ in dry acetonitrile, when cycled in the range from -0.07 to 1.07 V versus Fc/Fc$^+$ (being Fc = ferrocene). Reprinted from 'High performance Prussian Blue cathode for non-aqueous Ca-ion intercalation battery', Ref. [102], with permission from Elsevier

So far, the best results reporting the utilization of PB as a cathode for non-aqueous Ca-ion batteries were reached in a 3-electrode configuration, using a graphite rod as counter electrode and Ag/AgNO$_3$ as reference [102]. In 1M Ca(ClO$_4$)$_2$/dry acetonitrile, PB showed capacities of 94–118, 95 and 68–74 mAh·g^{-1} respectively at 18, 36 and 72 mA·g^{-1} (Fig. 4.17a). Curiously, the material got activated during the high rate tests, exhibiting 173 mAh·g^{-1} when returning to 18 mA·g^{-1} that quickly faded to 150 mAh·g^{-1}. By increasing the upper cut-off voltage from 1.0 to 1.37 V, 120 mA·g^{-1} were achieved at 125 mA·g^{-1}, of which 103 mA·g^{-1} were retained elapsed 80 cycles (Fig. 4.17b). This high capacity and cycle life were certainly unexpected, given the results above discussed, although we would like to highlight that the contribution of K$^+$ ions (initially present in PB) and its 'activating' role cannot be completely ruled out.

4.3.4 Prussian Blue as Cathode for Al-ion Batteries

Even the reversible intercalation of trivalent ions into PBA has been reported in organic electrolyte. To the best of our knowledge, Reed and coworkers were the first to cycle Cu-HCF using an aluminium-salt-based electrolyte (aluminium triflate dissolved in diglyme) [103]. Cu-HCF was active exclusively in highly concentrated electrolytes, exhibiting an initial charge capacity of 60 mAh·g^{-1} (corresponding to the extraction of K$^+$ ions) but a first discharge capacity that varied between 5 and 14 mAh·g^{-1} depending on cell. Such low reversible capacity was assigned to the intercalation of an aluminium-glyme complex (Al(Di)$_2^{3+}$) rather than bare Al^{3+}, at the [Fe(CN)$_6$]$^{4-}$ vacancies. Be that as it may, and despite the fact that the intercalation of Al^{3+} was very small, this work represents a starting point for the investigations of PBA in rechargeable Al-ion batteries.

4.4 Concluding remarks

The polyvalence of PBA/BGA/PWA has been manifested within this chapter and the previous one. There are still certain challenges to overcome, such as to improve the coulombic efficiency, even though big steps in this direction have been carried out by synthesizing materials with low vacancy and water content and high alkali content. Nevertheless, they have demonstrated their ability to act as cathodes in non-aqueous rocking chair batteries and their suitability to be adopted in several battery chemistries. In general, their good capacity (providing specific capacities as high as 160–210 mAh·g^{-1}), C-rate capability and cyclability, as well as their easiness of synthesis and low cost, position them as firm candidates for real applications. Proof of that is the investment that strong and consolidated companies, such as Sharp and governmental agencies (Arpa-E, i.e. Advanced Research Projects Agency-Energy), have executed.

References

1. N. Imanishi, T. Morikawa, J. Kondo, Y. Takeda, O. Yamamoto, N. Kinugasa, T. Yamagishi, J. Power Sources **79**, 215–219 (1999)
2. G. Zhuang, P.N. Ross, F.-P. Kong, F. McLarnon, J. Electrochem. Soc. **145**(1), 159–164 (1998)
3. S.-H. Yu, M. Shokouhimehr, T. Hyeon, Y.-E. Sung, ECS Electrochem. Lett. **2**(4), A39–A41 (2013)
4. L. Shen, Z. Wang, L. Chen, Chem. Eur. J. **20**, 1–5 (2014)
5. A.A. Karyakin, Electroanalysis **13**, 813–819 (2001)
6. R.D. Shannon, Acta Cryst. **A32**, 751 (1976)
7. X. Wu, M. Shao, C. Wu, J. Qian, Y. Cao, X. Ai, H. Yang, ACS Appl. Mater. Interfaces **8**, 23706–23712 (2016)
8. N. Imanishi, T. Morikawa, J. Kondo, R. Yamane, Y. Takeda, O. Yamamoto, H. Sakaebe, M. Tabuchi, J. Power Sources **81–82**, 530–534 (1999)
9. M. Okubo, D. Asakura, Y. Mizuno, J.-D. Kim, T. Mizokawa, I. Honma, J. Phys. Chem. Lett. **1**, 2063–2071 (2010)
10. D. Asakura, M. Okubo, Y. Mizuno, T. Kudo, H. Zhou, K. Amemiya, F.M.F. de Groot, J.-L. Chen, W.-C. Wang, P.-A. Glans, C. Chang, J. Guo, I. Honma, Phys. Rev. B **84**, 045117 (2011)
11. Y. Mizuno, M. Okubo, D. Asakura, T. Saito, E. Hosono, Y. Saito, K. Oh-ishi, T. Kudo, H. Zhou, Electrochim. Acta **63**, 139–145 (2012)
12. P.B. Bruce, M.Y. Saidi, J. Electroanal. Chem. **322**, 93 (1992)
13. P.G. Bruce, M.Y. Saidi, Solid State Ionics **51**, 187 (1992)
14. Y. Nanba, D. Asakura, M. Okubo, Y. Mizuno, T. Kudo, H. Zhou, K. Amemiya, J. Guo, K. Okada, J. Phys. Chem. C **116**, 24896–24901 (2012)
15. C.H. Li, Y. Nanba, D. Asakura, M. Okubo, D.R. Talham, RSC Adv. **4**, 24955 (2014)
16. M. Omarova, A. Koishybay, N. Yesibolati, A. Mentbayeva, N. Umirov, K. Ismailov, D. Adair, M.-R. Babaa, I. Kurmanbayeva, Z. Bakenov, Echim. Acta **184**, 59–63 (2015)
17. M. Okubo, D. Asakura, Y. Mizuno, T. Kudo, H. Zhou, A. Okazawa, N. Kojima, K. Ikedo, T. Mizokawa, I. Honma, Angewandte Chemie Int. Ed. **50**, 6269–6273 (2011)
18. D. Asakura, C.H. Li, Y. Mizuno, M. Okubo, H. Zhow, D.R. Talham, J. Am. Chem. Soc. **135**, 2793–2799 (2013)
19. M. Okubo, I. Honma, Dalton Trans. **42**, 15881 (2013)

20. D. Asakura, M. Okubo, T. Kudo, H. Zhou, T. Mizokawa, A. Okazawa, N. Kojima. Ext. Abstract (65th Fall Meeting, 2010), Phys. Soc. Jpn. 25aPS-57
21. D. Asakura, M. Okubo, Y. Mizuno, T. Kudo, H. Zhou, K. Ikedo, T. Mizokawa, A. Okazawa, N. Kojima, J. Phys. Chem. C **116**, 8364–8369 (2012)
22. C. Keng-Che, K. Ji-Jung, C. Fu-Rong, Electrochim. Acta **52**, 6554–6560 (2007)
23. S. Yagi, M. Fukuda, R. Makiura, T. Ichitsubo, E. Matsubara, J. Mater. Chem. A **2**, 8041 (2014)
24. K. Itaya, H. Akahoshi, S. Toshima, J. Electrochem. Soc. **129**(7), 1498 (1982)
25. T. Matsuda, Y. Moritomo, Appl. Phys. Express **4**, 047101 (2011)
26. Y. Moritomo, X. Zhu, M. Takachi, T. Matsuda, Jpn. J. Appl. Phys. **51**, 107301 (2012)
27. Y. Kurihara, T. Matsuda, Y. Moritomo, Jpn. J. Appl. Phys. **52**, 017301 (2013)
28. Y. Moritomo, M. Takachi, Y. Kurihara, T. Matsuda, Appl. Phys. Express **5**, 041801 (2012)
29. T. Matsuda, Y. Moritomo, J. Nanotechnol. Article ID 568147 (2012)
30. Y. Mizuno, M. Okubo, E. Hosono, T. Kudo, H. Zhou, K. Oh-ishi, J. Phys. Chem. C **117**(21), 10877–10882 (2013)
31. M. Shokouhimehr, S.-H. Yu, D.-C. Lee, D. Ling, T. Hyeon, Y.-E. Sung, Nanosci. Nanotechnol. Lett. **5**, 770 (2013)
32. P. Nie, L. Shen, H. Luo, B. Ding, G. Xu, J. Wang, X. Zhang, J. Mater. Chem. A **2**, 5852 (2014)
33. P. Xiong, G. Zeng, L. Zeng, M. Wei, Dalton Trans. **44**(38), 16746–16751 (2015)
34. M.J. Piernas-Muñoz, E. Castillo-Martínez, V. Roddatis, M. Armand, T. Rojo, J. Power Sources **271**, 489–496 (2014)
35. M.J. Piernas-Muñoz, E. Castillo-Martínez, E. Goikolea, P. Blanco, E. Legarra, J.S. Garitao-nandia, T. Fister, S. Kim, C. Johnson, T. Rojo, *Manuscript Under Preparation*
36. X. Sun, X.-Y. Ji, Y.-T. Zhou, Y. Shao, Y. Zang, Z.-Y. Wen, C.-H. Chen, J. Power Sources **314**, 35–38 (2016)
37. T. Shibata, M. Takachi, Y. Moritomo, Batteries **3**, 7 (2017)
38. M.N. Obrovac, V.L. Chevrier, Chem. Rev. **114**(23), 11444–11502 (2014)
39. X. Su, Q. Wu, J. Li, X. Xiao, A. Lott, W. Lu, B.W. Sheldon, J. Wu, Adv. Energy Mater. **4**(1), 1300882 (2014)
40. S.-W. Kim, D.-H. Seo, X. Ma, G. Ceder, K. Kang, Adv. Energy Mater. **2**, 710–721 (2012)
41. N. Yabuuchi, K. Kubota, M. Dahbi, S. Komaba, Chem. Rev. **114**, 11636–11682 (2014)
42. Y. Lu, L. Wang, J. Cheng, J.B. Goodenough, Chem. Commun. **48**, 6544–6546 (2012)
43. H. Lee, Y. Kim, J.-K. Park, J.W. Choi, Chem. Commun. **48**, 8416–8418 (2012)
44. M. Xie, Y. Huang, M. Xu, R. Chen, X. Zhang, L. Li, F. Wu, J. Power Sources **302**, 7–12 (2016)
45. H. Minowa, Y. Yui, Y. Ono, M. Hayashi, K. Hayashi, R. Kobayashi, K.I. Takahashi, Solid State Ionics **262**, 216–219 (2014)
46. H. Sun, H. Sun, W. Wang, H. Jiao, S. Jiao, RSC Adv. **4**, 42991 (2014)
47. R. Chen, Y. Huang, M. Xie, Q. Zhang, X. Zhang, L. Li, F. Wu, ACS Appl. Mater. Interfaces **8**, 16078–16086 (2016)
48. Y. Yue, A.J. Binder, B. Guo, Z. Zhang, Z.-A. Qiao, C. Tian, S. Dai, Angewandte Chemie Int. Ed. **53**, 1–5 (2014)
49. Y. You, X.-L. Wu, Y.-X. Yin, Y. Guo-Guo, Energy Environ. Sci. **7**, 1643–1647 (2014)
50. S.J.R. Prabakar, J. Jeong, M. Pyo, RSC Adv. **5**, 37545–37552 (2015)
51. J. Luo, S. Sun, J. Peng, B. Liu, Y. Huang, K. Wang, Q. Zhang, Y. Li, Y. Lin, Y. Liu, Y. Qiu, Q. Li, J. Han, Y. Huang, ACS Appl. Mater. Interfaces **9**(30), 25317–25322 (2017)
52. Y. Jiang, S. Yu, B. Wang, Y. Li, W. Sun, Y. Lu, M. Yan, B. Song, S. Dou, Adv. Funct. Mater. **26**(29), 5315–5321 (2016)
53. D. Yang, J. Xu, X.-Z. Liao, H. Wang, Y.-S. He, Z.-F. Ma, Chem. Commun. **51**(38), 8181–8184 (2015)
54. S. Yu, Y. Li, Y. Lu, B. Xu, Q. Wang, M. Yan, Y. Jiang, J. Power Sources **275**, 45–49 (2015)
55. H. Fu, C. Liu, C. Zhang, W. Ma, K. Wang, Z. Li, X. Lu, G. Cao, J. Mater. Chem. A **5**, 9604–9610 (2017)
56. Y. You, X.-L. Wu, Y.-X. Yin, Y. Guo-Guo, J. Mater. Chem. A **1**, 14061 (2013)
57. M. Okubo, C.H. Li, D.R. Talham, Chem. Commun. **50**, 1353 (2014)

58. M.J. Piernas-Muñoz, E. Castillo-Martínez, J.L. Gómez-Camer, T. Rojo, Electrochim. Acta **200**, 123–130 (2016)
59. X. Wu, W. Deng, J. Qian, Y. Cao, X. Ai, H. Yang, J. Mater. Chem. A **1**, 10130 (2013)
60. J.C. Pramudita, S. Schmid, T. Godfrey, T. Whittle, M. Alam, T. Hanley, H.E.A. Brand, N. Sharma, Phys. Chem. Chem. Phys. **16**, 24178–24187 (2014)
61. H. Wang, L. Wang, S. Chen, G. Li, J. Quan, E. Xu, L. Song, Y. Jiang, J. Mater. Chem. A **5**, 3569 (2017)
62. S. Kajiyama, Y. Mizuno, M. Okubo, R. Kurono, S. Nishimura, A. Yamada, Inorg. Chem. **53**, 3141–3147 (2014)
63. L. Wang, Y. Lu, J. Liu, M. Xu, J. Cheng, D. Zhang, J.B. Goodenough, Angewandte Chemie Int. Ed. **52**, 1–5 (2013)
64. J. Song, L. Wang, Y. Lu, B. Guo, P. Xiao, J.-J. Lee, X.-Q. Yang, G. Henkelman, J.B. Goodenough, J. Am. Chem. Soc. **137**(7), 2658–2664 (2015)
65. D.A. Stevens, J.R. Dahn, J. Electrochem. Soc. **147**(4), 1271–1273 (2000)
66. Y. You, X.-Q. Yu, Y.-X. Yin, K.-W. Nam, Y.-G. Guo, Nano Res. **8**(1), 117–128 (2015)
67. Y. Liu, Y. Qiao, W. Zhang, Z. Li, X. Ji, L. Miao, L. Yuan, X. Hu, Y. Huang, Nano Energy **12**, 386–393 (2015)
68. L. Wang, J. Song, R. Qiao, L.A. Wray, M.A. Hossain, Y.-D. Chuang, W. Yang, Y. Lu, D. Evans, J.-J. Lee, S. Vail, X. Zhao, M. NIshijima, S. Kakimoto, J.B. Goodenough, J. Am. Chem. Soc. **137**(7), 2548–2554 (2015)
69. W.-J. Li, S.-L. Chou, J.-Z. Wang, Y.-M. Wang, Y.-M. Kang, J.-L. Wang, Y. Liu, Q.-F. Gu, H.-K. Liu, S.-X. Dou, Chem. Mater. **27**, 1997–2003 (2015)
70. X. Yan, Y. Yang, E. Liu, L. Sun, H. Wang, X.-Z. Liao, Y. He, Z.-F. Ma, Electrochim. Acta **225**, 235–242 (2017)
71. Y. Liu, G. Wei, M. Ma, Y. Qiao, Chem. Eur. J. **23**, 15991–15996 (2017)
72. W. Ren, M. Qin, Z. Zhu, M. Yan, Q. Li, L. Zhang, D. Liu, L. Mai, Nano Lett. **17**, 4713–4718 (2017)
73. M.J. Piernas-Muñoz, E. Castillo-Martínez, O. Bondarchuk, M. Armand, T. Rojo, J. Power Sources **324**, 773–776 (2016)
74. J.-Y. Liao, Q. Hu, B.-K. Zou, J.-X. Xiang, C.-H. Chen, Electrochim. Acta **220**, 114–121 (2016)
75. Y. Liu, D. He, R. Han, G. Wei, Y. Qiao, Chem. Commun. **53**, 5569–5572 (2017)
76. X. Wu, C. Wu, C. Wei, L. Hu, J. Qian, Y. Cao, X. Ai, J. Wang, H. Yang, ACS Appl. Mater. Interfaces **8**(8), 5393–5399 (2016)
77. M. Takachi, T. Matsuda, Y. Moritomo, Appl. Phys. Express **6**, 025802 (2013)
78. M. Takachi, T. Matsuda, Y. Moritomo, Jpn. J. Appl. Phys. **52**, 090202 (2013)
79. T. Matsuda, M. Takachi, Y. Moritomo, Chem. Commun. **49**, 2750 (2013)
80. D. Yang, J. Xu, X.-Z. Liao, Y.-S. He, H. Liu, Z.-F. Ma, Chem. Commun. **50**, 13377–13380 (2014)
81. M. Xie, M. Xu, Y. Huang, R. Chen, X. Zhang, L. Li, F. Wu, Electrochem. Commun. **59**, 91–94 (2015)
82. H.-W. Lee, R.Y. Wang, M. Pasta, S.W. Lee, N. Liu, Y. Cui. Nat. Commun. (2014), https://doi.org/10.1038/ncomms6280
83. Y. Xu, S. Zheng, H. Tang, X. Guo, H. Xue, H. Pang, Energy Storage Mater. **9**, 11–30 (2017)
84. A. Paolella, C. Faure, V. Timoshevskii, S. Marras, G. Bertoni, A. Guerfi, A. Vijh, M. Armand, K. Zaghib, J. Mater. Chem. A **5**, 18919 (2017)
85. J. Qian, C. Wu, Y. Cao, Z. Ma, Y. Huang, X. Ai, H. Yang, Adv. Energy Mater. 1702619 (2018)
86. A. Eftekhari, Z. Jian, X. Ji, ACS Appl. Mater. Interfaces **9**(5), 4404–4419 (2017)
87. A. Eftekhari, J. Power Sources **126**, 221–228 (2004)
88. C. Zhang, Y. Xu, L. Liang, H. Dong, M. Wu, Y. Yang, Y. Lei, Adv. Funct. Mater. 1604307 (2017)
89. Z. Shadike, D.-R. Shi, Tian-Wang, M.-H. Cao, S.-F. Yang, J. Chen, Z.-W. Fu, J. Mater. Chem. A **5**, 6393–6398 (2017)
90. X. Wu, Z. Jian, Z. Li, Z. Ji, Electrochem. Commun. **77**, 54–57 (2017)

91. L. Xue, Y. Li, H. Gao, W. Zhou, Z. Lü, W. Kaveevivitchai, A. Manthiram, J.B. Goodenough, J. Am. Chem. Soc. **139**, 2164–2167 (2017)
92. G. He, L.F. Nazar, ACS Energy Lett. **2**, 112–1127 (2017)
93. C. Ling, J. Chen, F. Mizuno, J. Phys. Chem. C **117**, 21158–21165 (2013)
94. R.C. Massé, E. Uchaker, G. Cao, Sci. China Mater. **58**(9), 715–766 (2015)
95. X. Sun, V. Duffort, L.F. Nazar, Adv. Sci. 1600044 (2016)
96. W.M. Hayes (ed.), *CRC Handbook of Chemistry and Physics*, 95th edn. (CRC Press, 2014)
97. P. Padigi, G. Goncher, D. Evans, R. Sonalki, J. Power Sources **273**, 460–464 (2015)
98. A. Ponrouch, C. Frontera, F. Bardé, M.R. Palacín, Nat. Mater. **15**, 169–172 (2016)
99. T. Shiga, H. Kondo, Y. Kato, M. Inoue, J. Phys. Chem. C **119**, 27946–27953 (2015)
100. A.L. Lipson, B. Pan, S.H. Lapidus, C. Liao, J.T. Vaughey, B.J. Ingram, Chem. Mater. **27**, 8442–8447 (2015)
101. T. Tojo, Y. Sugiura, R. Inada, Y. Sakurai, Electrochim. Acta **207**, 22–27 (2016)
102. N. Kuperman, P. Padigi, G. Goncher, D. Evans, J. Thiebes, R. Solanki, J. Power Sources **342**, 414–418 (2017)
103. L.D. Reed, S.N. Ortiz, M. Xiong, E.J. Menke, Chem. Commun. **51**, 14397–14400 (2015)

Chapter 5
Conclusions and Perspectives

5.1 Introduction

From the origins of the Electrochemistry back in the eighteenth century, a huge progress has been achieved in this field, especially in the topic of batteries.

Briefly, a battery is a device capable of converting the chemical energy contained in the active materials that compose it into electric energy by redox reactions. Primary and secondary batteries have served for several applications along these centuries, but the battery revolution has undoubtedly emerged via utilization of Li-ion batteries in portable electronic devices and more recently in electric cars. This fact has triggered the investigation of new battery technologies, such as Li–air, Li–sulphur and A-ion batteries (A = Na, K, Mg, Ca, Zn, Al) with the purpose to overcome the limitations that Li-ion batteries present, either in terms of electrochemical properties or material abundance.

5.2 Overview on A-ion Batteries

A plethora of materials have been investigated and new ones continue to be proposed for their applications as cathodes and anodes for A-ion batteries. However, these vary from technology to technology.

For Li-ion batteries, the cathodes more commonly studied are layered oxides, spinel-type oxides and polyanionic compounds [1]. The anode par excellence is graphite, although silicon and other alloy-type material are outlined for high-energy density applications. Titanates and conversion materials, such as oxides, sulfides and fluorides, have also been investigated [2].

© The Author(s) 2018
M. J. Piernas Muñoz and E. Castillo Martínez, *Prussian Blue Based Batteries*,
SpringerBriefs in Applied Sciences and Technology,
https://doi.org/10.1007/978-3-319-91488-6_5

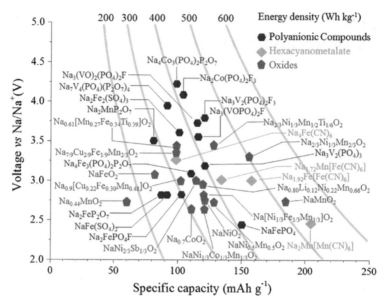

Fig. 5.1 Voltage versus capacity including representative examples of the most studied families of cathode materials for Na-ion batteries Reproduced from Ref. [3] with permission from Wiley

Despite Na-ion batteries were initially investigated in parallel with LIB, the latter took soon the lead and the R & D of the former was not resumed until 2011. Nonetheless, the enormous efforts invested in the NIB technology within the last 7 years have enabled its development at a dizzying pace. The materials originally explored as cathodes in NIB [3] are mainly analogues to those employed in LIB—layered oxides and polyanionic compounds (see Fig. 5.1). Although, since bigger sites are required to allocate Na^+, Prussian Blue and Analogue materials with larger cavities have arisen as a strong alternative. As for the anodes, hard carbon [4] is so far the preferred material to work with and has been implemented in full cell prototypes.

K-ion batteries are also receiving great attention nowadays. The possible intercalation of K^+ into graphite has been demonstrated, thus bringing this type of batteries one step closer to the achievements of LIB [5]. Among the cathode materials, layered oxides (that worked in precedent A-ion technologies) present, here, important drawbacks, as they undergo severe structural changes by insertion/extraction of such big cation (K^+). Conversely, Prussian Blue Analogues' channel size perfectly matches K^+ and are postulated as one of the most promising cathode candidates for real KIB applications [6].

On the other hand, Mg-, Ca- and Al-ion batteries are still at early stages of research and suffer from several problems, as slower mobility of the A^{n+} (n = 2 or 3) ions derived from their higher formal charge and the necessity of finding appropriate electrolyte and electrode materials.

5.3 Prussian Blue Materials

The simplicity of synthesis, the abundance of elements that generally comprise them and the 3D metalorganic framework with open channels, make Prussian Blue and their analogues appealing materials. This has fostered an intense research towards different applications, being electrochemical energy storage one of the utmost importance.

The structure of Prussian Blue can be described as a 3D cubic network composed by iron atoms (ferrous (Fe^{II}) and ferric (Fe^{3+})) alternatively located at the corners of a small cube of 5.1 Å edge linked by bidentate $-C\equiv N-$ ligands, featuring zeolitic sites capable to host species with an ionic radius ≤ 1.6 Å. Depending on the synthesis and the reactants used, PB materials with compositions between those of '*insoluble*' Prussian Blue ($Fe^{3+}_4[Fe^{II}(CN)_6]_3 \cdot xH_2O$, which contains one-fourth of $[Fe^{II}(CN)_6]^{4-}$ vacancies and no alkali cation in its structure) and '*soluble*' Prussian Blue ($AFe^{3+}[Fe^{II}(CN)_6 \cdot xH_2O$, that allocate alkali cations but it is vacancy free) are produced.

Two more species closely related to '*soluble*' Prussian Blue include the so-called *Berlin Green*, $Fe^{3+}[Fe^{III}(CN)_6]$, and *Prussian White*, $A_2Fe^{2+}[Fe^{II}(CN)_6]$, which can be defined as the totally oxidized and reduced phase of Prussian Blue. Both can be obtained by via electrochemical process or direct synthesis (as a function of A). With respect to the structure, Berlin Green preserves the cubic symmetry, while Prussian White can crystallize in cubic structure or in lower symmetry, either monoclinic or rhombohedral depending on the alkali cation occupying the zeolitic cavities.

Substituting 'Fe' by other transition metals Prussian Blue Analogues (PBA) of formula $AM[M'(CN)_6] \cdot xH_2O$ (being usually $A =$ alkali; M, $M' =$ transition metal; typically $M' =$ Fe) are obtained. PBA typically adopts a cubic structure analogue to PB, though there are some exceptions. Similarly, Berlin Green Analogue ($M^{3+}-N\equiv C-M'^{III}$) and Prussian White Analogue ($M^{2+}-N\equiv C-M'^{II}$) phases can also be prepared. Their symmetry again depends on the transition metals and, for PWA, on the alkali or alkali-earth present in the structure.

This book has been mostly devoted to review the electrochemical performance of this class of materials (PB and PBA) in batteries. A glance at the most relevant aspects extracted from it is given below.

5.4 Key Aspects of the Application of Prussian Blue Materials to Batteries

Aqueous batteries. We have seen that the electrochemistry of PB and analogues was first explored in aqueous media, using thin films produced by electrodeposition. Among others, those studies enabled: (i) observing the colour change upon reduction and oxidation and establishing the fast electrochromism of PB; (ii) performing fundamental studies on the thermodynamics of the reaction and determining that only alkali cations and not H^+ were inserted into the cavities of the PB structure; (iii) revealing the oxidation of Fe^{II}–C at high voltage and the reduction of ionic Fe^{3+}–N at lower voltages by Mössbauer spectroscopy studies and (iv) understanding the influence of each of the metals, M and M', in the voltages of reaction. Nevertheless, electrodeposition mostly led to materials of composition of the PB type, and control over the stoichiometry, $[Fe(CN)_6]^{4-}$ vacancies and alkali content seem difficult.

Recent studies on bulk electrodes of PB and PBA have shown the implications that the use of one metal or another has on the electrochemical properties of these materials. Ni–HCF and Cu–HCF, for instance, store 1 alkali (or half alkali-earth) ion and 1 e^- per f. u. (~60 mAh g^{-1}) with a minimal lattice strain of ca. 1% of the cubic lattice parameter. Among PBA, these present the fastest kinetics and longest cycle life (10^4 cycles). Moreover, and likely linked to their good structural stability, they have also shown the insertion of divalent and trivalent cations, which have preference to occupy the $[Fe(CN)_6]^{4-}$ vacancies. On the other hand, PB and PBA based on other transition metals (M=Mn, Fe, Co) generally experience redox reactions involving 2 e- per f.u., since the two metallic centers participate on them. However, although they can show much higher capacities than Ni- or Cu-HCF, typically 100–170 mAh g^{-1}, they often undergo structural transitions which result in more sluggish kinetics and poorer cycle life.

Some research on low cost and environmentally friendly aqueous full cells has also been conducted. At lab scale, full cells capable of operating at ca. 1 V, and reaching cycle lives of up to 10^4 cycles, have been built with a PBA cathode and carbon or organic anodes. Using Zn metal as anode, Zn–HCF as cathode and an electrolyte stabilized by highly concentrated salts and surfactants results in higher energy density (1.7 V) full cells. Still, further improvements would be needed to make these batteries efficient.

Non-aqueous batteries. In general, non-aqueous batteries present slower kinetics than aqueous-based batteries, mostly due to the need of the cations in organic solvent to be desolvated before entering the PB structure. Still, non-aqueous batteries based on PB and PBA have shown promising performance. For instance, in NIB, they have exhibited average voltages and specific capacities comparable and competitive to those attained by other proposed cathodes for this technology (Fig. 5.1), which adds new advantages aside the low cost of these materials.

Some of the key findings and lessons learned from the performance of PBA, BGA and PWA that can be extracted from the previous chapter are the following:

The suppression of $[Fe(CN)_6]^{4-}$ vacancies translates into the reduction of coordinated water and the supply of larger amount of redox active centers available for A^+ ion storage, enhancing the specific capacity and coulombic efficiency, but also resulting in better structural stability and thus greater capacity retention and rate capability. This vacancy suppression can be achieved by different synthetic methodologies, including those focused on slowing down the crystallization speed and those producing PB or PBA composites with highly conductive species of carbon (RGO, graphene rolls, etc.).

Transition metal substitution permits to tune the properties of PB, leading to PBA that show substantial changes in their voltage and capacities (as it occurs in aqueous batteries). Specifically, the presence of two active redox pairs enables reaching higher capacities. Conversely, total or partial replacement of redox active metals by inactive redox metals can enhance the cycle stability and, sometimes, further enlarge the capacity of the Fe–C redox pair. For instance, Mn–HCMn and Fe–HCF provide the largest capacities in NIB, while Ni–HCF behaves as a zero-strain material exhibiting one of the best cyclabilities.

The presence of one alkali or other in the pristine materials can modify the electrochemical properties (voltage, capacity and cyclability), affecting occasionally to the A^+ insertion/extraction mechanism, and influencing as well the activation energy of the interfacial (electrode-electrolyte) charge transfer process.

The synergistic effects of a dual cation insertion into PBA/PWA are beneficial for the electrochemical performance, either enlarging the cycle life of the material and increasing the voltage at which the redox reactions occur, as in the case of Na^+/K^+ insertion, or increasing the capacity, as with Li^+/Mg^{2+} or K^+/Ca^{2+} intercalations.

PBA have also proved their versatility, being capable of acting both as cathodes and anodes. A conversion reaction is suggested to happen at low voltages of operation of HCF anodes in LIB, whereas the mechanisms of HCMn and HCCo have not been clarified yet and could involve metal reduction below 2+ without experiencing conversion reaction. The few studies conducted on NIB, have, however, suggested little or no activity.

Last but not least, the contribution of other cell components, such as the electrolyte, binder or type of conductive carbon added to the mixture have also been evidenced to play a role, since different combinations lead to distinct characteristics. Although this applies to any battery type and it is not specific of Prussian Blue materials, it points out that a careful selection is necessary to reach optimal electrochemical properties.

5.5 Future Perspectives

PB and PBA materials, although known for centuries, still pose some unknowns while they keep showing up as promising materials for new applications in several research fields.

On the one hand, the synthetic methodologies to achieve specific compositions are still not fully established and they are just at the stage of being generalized from one stoichiometry to another. On the other hand, despite numerous attempts to study their crystal structure, many technologically important phases are nanosized so only the average crystal structure is known. With new recent developments, more insights into the atomic arrangements including vacancies, water and alkali short range orderings should be achieved. For instance, low dose electron microscopy should enable performing electron diffraction in typically beam sensitive materials, more advanced detectors for X-ray and neutron diffractometers would allow resolving smaller crystals and developments in synchrotron techniques could help us look at disordered materials.

From the perspective of their applications, two examples of possible short-term commercialization in rechargeable batteries are conceived, mainly for grid storage. These are Prussian Blue analogue phases with a single active redox center as cathodes for aqueous batteries and high-quality Prussian White phases as cathodes for organic batteries, especially for NIB and KIB. As well, aqueous Zn/PB batteries show promise for future deployment. Nonetheless, given the redox activity of these materials and the vast number of combinations that can be performed, their exploration in other fields of research related to electrochemistry such as electrochromic devices and electrocatalysis, or emerging areas such as artificial photosynthesis or CO_2 capture can also be foreseen.

We would like to finish this chapter and this book calling the attention of the reader to the fact that, regardless of the material employed, recycling or integrating batteries in a circular economy cycle is essential to ensure that they serve their purpose of contributing to a more environmentally friendly energy landscape. Otherwise, we will just be shifting the problem from contamination by fossil fuels to huge battery waste generation. Although with increasingly lower cost battery materials, the interest in recycling its components would decay unless government regulations and a responsible design avoid it; we would like to be optimistic and foresee that battery recycling will become an important issue and research topic in the upcoming years.

References

1. N. Nitta, F. Wu, J.T. Lee, G. Yushin, Materials Today **18**(5), 252 (2015)
2. G. Jeong, Y-U. Kim, H. Kim, Y-J. Kim, H-J. Sohn, Energy Environm. Sci. **4**, 1986 (2011)
3. C. Zhao, Y. Lu, Y. Li, L. Jiang, X. Rong, Y.-S. Hu, H. Li, L. Chen, Small Methods **5**, 1700063 (2017)
4. D.A. Stevens, J.R. Dahn, J. Electrochem. Soc. **147**(4), 1271–1273 (2000)
5. S. Komaba, T. Hasegawa, M. Dahbi, K. Kubota, Electrochem. Commun. **60**, 172 (2015)
6. A. Eftekhari, Z. Jian, Z. Ji, ACS Appl. Mater. Interfaces **9**(5), 4404 (2017)

Printed in the United States
By Bookmasters